FUNDAMENTALS OF AGILE PROJECT MANAGEMENT

An Overview

Marcus Goncalves and Raj Heda

The Technical Manager's Survival Guides

Library of Congress Cataloging-in-Publication Data

Gonтalves, Marcus.
  Fundamentals of agile project management : an overview / Marcus Goncalves and Raj Heda.
      p. cm.
  Includes bibliographical references and index.
  ISBN 978-0-7918-0296-0 (alk. paper)
  1. Project management. 2. Production management. 3. Flexible manufacturing systems. 4. System design. I. Heda, Raj. II. Title.
   T56.8.G645 2010
   658.4'04--dc22
                            2009048509

Table of Contents

Acknowledgement

Many thanks (again!) to Mary Grace Stefanchik, the editor at the American Society of Mechanical Engineers (ASME), for not only publishing yet another one of my titles for ASME's collection, but especially for her patience during the production phase. Many thanks to my co-author and friend Raj Heda, for finding time in his schedule at IBM and lecturing at Nichols College, to write land his expertise on Agile and write this book with me.

I would also like to express my appreciation to many corporate leaders that shared their views and experiences with us about Agile.

Many thanks also go to my spiritual partners at the Boston Church of Christ, Steve Major and Matt Paradise, for their continuous spiritual support and friendship. Last but not least, my deepest gratitude to my wife Carla, sons Samir and Josh (in memory), and my princess Andrea (also in memory), the true joy of my life.

Raj Heda: I wish to record my debt to some of the people who have made an indelible mark in my life.

A special note of thanks to my dear professor and now friend and colleague, Marcus, for his generous helpfulness, trust, support and above all, for offering me the opportunity to co-author this book.

To my mother for being ever loving and encouraging.

To my brother Ravi for all the love and the fun days we spent.

To my aunt, Meenu for always lending me a patient ear and giving me genuine advice in all my endeavors.

To my friend and colleague in business, Dorothy, for her sincere look-out for my well-being and for her beautiful heart.

To my dear friends Anand, Shrikant, Prajay, Prashant and Amit for always being there for me in good times and bad.

To my good friend, Matt, for help with graphics in the book. To my wife, Anu, for being such a caring and loving life partner, for her synergistic help in all my activities and for her invaluable editing of this book. To my beautiful princesses Radhika and Vrinda, who are my true loves and who make it all worth the while!

Dedication

To my forever beautiful wife Carla, my son Samir living here on earth, and my children Andrea and Joshua living in Heaven. To God be the glory!

Marcus Goncalves, Spring 2009

To my ever beautiful and loving Lord. To my spiritual grandfather, A.C. Bhaktivedanta Swami Srila Prabhupada and my spiritual guide, Niranjana Swami, for being a constant ray of light in my life. To the memory of my father, my dearest friend and my best guide.

Raj Heda, Spring 2009

Chapter 1
A Case for Agile

*The process of evolution can only be described as the
gradual insertion of more and more freedom into matter[1].*

Changing Landscape of Project Management

Project management (or PM) has been around for almost as
long as projects have been around; in other words, since time
immemorial. Of course, formalization of project management
terminologies and techniques is not quite as dated and is also
evolving as we speak. Project management has not achieved the
same level of maturity and acceptance in all geographies and
industries, neither is there one silver-bullet PM method that will work
in all situations. The methods, principles and practices used to
manage projects depend on various factors including the criticality,
complexity, and length of the project, as well as team size and
location, organizational culture, etc.

There is no PM practice that is universal. Besides, even for
the same domains, practices evolve with time and also in response to
unsuccessful attempts in the past. The Project Management Institute
estimates that the world spends nearly one-fifth of its GDP (US $12
trillion in 2008) on projects. Given the huge amounts of funds
involved, it is essential to gain better understanding of PM methods
and be able to apply them in the best possible manner.

In this book, we make an attempt to introduce the readers to
agile methods to managing projects. Agile methods have been
around for a while now. However, they are gaining more and more
acceptance as a result of the success trail that they leave behind in
organizations that apply them, and also as organizations that have
been burned by unsuccessful projects look for alternatives so that

---

[1] T.E. Hulme, " The Philosophy of Intensive Manifolds," Speculations:
Essays on Humanism and the Philosophy of Art, 1924

history does not repeat itself. Indeed, the yet to be released Standish Group report on software development chaos gives credit to agile methods for the stellar improvement in software development projects in the last decade. The new report states that 35% of software projects started in 2006 can be categorized as successful; a marked improvement from the groundbreaking report in 1994, which categorized only 16.2% projects as successful.

As per the Chairman of the Standish Group, Jim Johnson, the three primary reasons for the improvement in software quality are better project management, iterative development and the emerging Web infrastructure. While some of the agile methods discussed here spring from software development experiences, it is essential to understand that the underlying principles can be applied with appropriate modifications to any industry and organization.

Why Projects Fail

Imagine a war situation a few decades back when communication tools were not very advanced. A company of soldiers is sent out by the captain for a specific mission with a specific set of commands. The soldiers are expected to exactly follow the commands and have no discretionary powers whatsoever. They set out for the mission. While on field, the situation changes significantly, or the enemy acts in certain unforeseen ways. However, the soldiers have no facilities for two-way communication with the captain and are not sure if they should really bend the commands to deal with the situation in hand.

What do you think will be the outcome of the above mission? The company of soldiers will possibly see the mission through or might get hurt in the process. However, they might be completely disillusioned (or even worse) because they know the original commands are not helpful anymore, and they will certainly lack a sense of ownership for the mission. The mission might be abandoned well before its completion. Even if it does get completed, it might not have relevance any more. The captain might have to answer some very uncomfortable questions from higher-ranked officials.

The mission in the above example can be compared to a project handled by executives (captains) in various organizations and executed by project teams (company of soldiers). The example shows what traditional project management methods can do to a project in a changing business environment. With traditional methods, we could end up with a customer who invests in a project that serves no purpose, a project team that just works to complete the project and has no concern for the usefulness of the final product, and a final product that may very well be obsolete, or of such poor quality that it is effectively useless at the end of the project.

This leads us to the question - why is it that so many projects in the best companies handled by the best PMs fail, despite in-depth planning and documentation, very highly skilled project team and tons of dollars spent on them? This is a serious area of concern, particularly for software development projects. It is a known fact in IT circles that software projects are prone to greatly exceeding budget and schedule.

Some of the key reasons for project failure are:

- A fatal condition called "analysis paralysis;"
- Lack of stakeholder commitment and ownership;
- The project manager works on what he thinks the client wants rather than what he knows the client needs;
- Lack of collaboration and communication;
- Emphasis on documentation rather than working results;
- Companies focus on 'plan the work and work the plan' and leave no window of opportunity to get intermediate feedback from the stakeholders and incorporate it into the work;
- Insufficient focus on the business value that the project should produce.

The above list is in no way comprehensive but it does give us enough justification to look for solutions that increase the chances for successful project execution. In the current economy, where organizations are under immense pressure to turn out reliable projects on time and under budget, even as project budgets are being slashed across the board, organizations find themselves looking for better tools and PM methods to survive and thrive.

Case Example: Evolution of Software Development Methodologies

Figure 1.1 shows the evolution of software development methodologies in the last few decades. Since most of the agile methods have their roots and are applied to software development, we are tracing the history of software development methodologies. This evolution is not very sacrosanct.

You will find examples of many overlaps and methodologies used much before the decade in which they are presented in the figure. The figure is just to give the reader an idea of which methodology gained prominence and got formalized in which decade. The figure does show that we have a come a long way from the linear life cycle or waterfall thinking of the 60's, that required dotting every i and crossing every t of the project plan before moving on to execution.

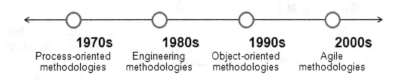

1970s	1980s	1990s	2000s
Process-oriented methodologies	Engineering methodologies	Object-oriented methodologies	Agile methodologies

Figure 1.1 – Methodology trends through the years

Is "Work the Plan and Plan the Work" Obsolete?

Not so long ago, it was widely accepted that planning is a very critical part of the entire project. While proper planning is still very important and will result in better execution, the timeframe for the project plan is shrinking. In the current environment, it is very difficult to plan for the entire project life cycle at the start of the project. There is a need to make frequent visits to the drawing board. It is quite possible that the project scope may change entirely during the various iterations of project planning; however, this project may still be classified as successful, while the project that delivers what was planned for but became obsolete during project execution may be classified as redundant or a failure.

In today's environment, it may not be wrong to say that all products and services are perishable. If the time to market is too long, the product or service may be rendered useless even if it has not physically perished. In the same way, a project can very easily perish if by the time it is completed, the stakeholder demands have changed. Companies have begun to realize that people and not processes are the real lubricants for any project and hence stakeholder commitment and involvement is very necessary for the success of projects. In order for the stakeholders to be committed to the project, they must be able to see its usefulness in their current environment and lives.

As a result, in the current dynamic and uncertain world, traditional project management methods are getting less and less effective and this calls for methods that are lot more dynamic, adaptive, flexible…. in other words, agile.

Assembling a Jigsaw Puzzle

Agile, in simple terms can be defined as an iterative and incremental approach to managing projects, with every iteration delivering a complete, working subset of the final product.

Agile projects can be compared to assembling a jigsaw puzzle. Every piece is an increment that brings us closer to putting together the whole picture. Let us take a moment to think about how we solve a puzzle. It consists of a sequence of steps that include: deciding which pieces form the core of the picture or will help us in putting together the rest; taking that piece out of the crowd, placing it on the board or integrating it with other pieces; taking a moment to analyze if the piece belongs where it was put; and then going back and starting the sequence over again.

In a similar fashion agile project management (or APM) consists of executing a project in iterations. Each iteration is a set of steps comprised of requirements analysis, design, development, testing, implementation and integration. Each iteration delivers a complete component of the final product, which is an increment beyond the previous iteration. A sequence of such incremental iterations delivers the final product.

Agile working is based on skilled improvisation in place of extensive planning and documentation. It is difficult to plan for all contingencies and to plan for an external environment that is beyond our control and where the only certainty is change. In such an environment, we need methods that are flexible enough to permit switching gears without stalling the whole process. The stage is set for agile. In the agile methodology, change is not just expected but is readily embraced. Agile is just adaptive and not predictive.

Perfect communication is a very difficult, if not impossible, concept to achieve. Hence, perfect understanding of the project requirements is very difficult. Not only is it very difficult to fully understand what the customer wants from the final product, quite often, even the customer doesn't really know what he wants and continues to polish his requirements as the project proceeds and he gets to test the initial iterations. This is acceptable in agile project management. In the traditional methods, the requirements get frozen quite early in the project and there is little, if any, scope for changes thereafter. Agile presents a significant shift in project management by allowing design to change while product is being developed. To the old school thinkers, this might look like inviting chaos; but agile is actually an attempt at contained chaos.

When we say that agile methodologies readily embrace change, it is not to be believed that agile project requirements are forever changing and evolving. More often than not, the customer is unable to distinctly define project requirements at the start of the project. However, after the first few iterations, the customer is more clear about his expectations from the final product and the requirements become more distinctly defined.

It can thus be summarized that traditional methods work best when the final outcome is well-defined and the process is repetitive (example mass production of commodities). Agile methods are useful when the final outcome is not certain, when what is desired is possibly known but the way to get there is uncertain and more than one ways to get there is acceptable. Thus, agile is applicable in areas where a degree of inexactness and imprecision is acceptable. It thrives in projects where uncertainty is the name of the game, and where lateral and out-of-the-box thinking are welcomed.

While most of the agile methods have evolved from software development experiences, some of the agile practices take inspiration from lean manufacturing principles and the Theory of Constraints. Besides, there are industries that have been using the agile methods, seemingly always, even without knowing it.

A classic example of this is the healthcare industry. If a patient's health condition is rather complicated (like a complex project) the doctor requires the patient to make multiple visits (just like iterations in a project). A visit to an emergency room of any hospital will also be sufficient to see smooth agile project management at work. The patients in the ER are prioritized based on the criticality of their situation, quite similar to the prioritizing of tasks based on value generation capability in projects. Of course, there are other aspects where healthcare totally diverges from APM. For instance, the healthcare industry's focus on comprehensive documentation is different from APM's focus on working results.

The nature of certain projects might be such that detailed upfront planning is required. Consider the example of a road construction project in a third-world country funded by the World Bank. Such projects usually suffer from a multitude of problems including lack of shared goals between the project sponsor, local government and the general public affected, as well as unclear responsibilities and bureaucratic delays. Another major concern is that such projects last for such a long period that they end up being administered under more than one government. This often causes further delays and a political "blame game" situation. Such projects will require comprehensive planning at the start of the project and more than one individual will be responsible for work on any particular item on the project task list. There will be a requirement for more formal and detailed reporting.

One very relevant question in this context is "Can APM principles be applied in such instances?" The answer is a resounding "Yes." It is possible that APM can not be applied *in toto;* however, various APM practices can be applied as relevant. Indeed APM can possibly be the answer to the deep-set malaise of cost and time overruns in such projects. With the agile, every government in its turn will witness some deliverables, and hence can take responsibility for them. The project cost will not be as daunting with quicker deliverables, and APM will provide the flexibility to project sponsors to cut funding if the expected results do not materialize.

Understandably, the deliverable in such projects will not be in implementable stage until much later into the project. But as we said, the agile principles can be applied as relevant, and it need not be an all or nothing situation for industries wishing to adopt APM.

Do We Still Need Project Managers?

In agile methods, the primary focus seems to have shifted to the project team. The agile development team is self-organizing and works under minimal control. Does this mean that there is no need for project managers under agile methods?

Indeed, under agile project management, the role of a project manager has undergone significant change. With agile methodologies, project management becomes an art. No number of certifications for project management will suffice to make one a successful agile project manager.

Project managers using the agile methods can be compared to backpackers. Such managers manage a project without much baggage of contracts, planning and documentation. As such, they are more open to change based on changing needs and environment.

This however, puts additional responsibility on the manager to be up to date with the changing needs of the stakeholders.

Agile is not for project managers who work like military generals. Agile project managers have to be willing to relinquish control to the team and must trust the team members. Agile project managers do not really manage or organize the project tasks, they just act as facilitators. Agile methods put a high premium on the soft skills of the manager. The manager has to be a good communicator. He has to work to enhance the team performance by removing impediments from its way while keeping the business vision in mind. In short, an agile project manager has to adaptively lead his team and not micromanage it based on elaborate plans and documents.

Agile Benefits

Different organizations have different tracks of project outcomes and hence different reasons for adopting agile methods. Figure 1.2 summarizes some of the reasons why many software organizations are going the agile way.

• Accelerate time to market	22
• Enhance Ability to manage changing priorities	21
• Increase productivity	12
• Enhance software quality	10
• Improved alignment between IT and business	9
• Improve project visibility	6
• Reduce risk	6
• Simplify development process	4
• Other	3
• Improved/increased engineering discipline	2
• Enhance software maintainability/extensibility	2
• Improved team morale	1

Figure 1.2 – Most important reasons for adopting Agile by percentage – Source: VersionOne

Some might question the effectiveness of agile, and have concerns that with less planning and documentation, the project is going to be in a constant state of chaos and change. While this is not true, it does make us consider whether a little chaos is better than a defunct final product.

Of course, product developed agilely is not a guarantee for an innovative or successful product; however, it does increase the chances of delivering a better product, or a product that more closely meets customer's evolving requirements.

Agile has several benefits over traditional project management, which provide enough reasons for organizations to sit up, take notice, and adopt or adapt, as the case may be.

- *Adaptability:* With agile methods, the project is not front-loaded but the weight is equally distributed throughout the various phases of the project. Requirements gathering and planning are ongoing processes in agile projects and not things that get completed in the first phase of the project. What this implies is that the project goals need not be frozen at the start of the project but can be adapted based on the changing customer needs. What this also implies is that product design, development and all the subsequent steps need not wait in line until a comprehensive list of project requirements is built.

- *Promotes innovation:* Organizations that reward and encourage out-of-the-box thinking and risk-taking are the biggest proponents of agile methods. In such organizations, the think tanks are at work even during project execution. They are not just thinking about how to complete the task at hand, but also about how to enhance the effectiveness of the final output - keeping in mind the business value goal of the project. This does not mean that the project is just allowed to go on a tangent, or that it goes into a never-ending iterative loop. However, innovation within reasonable limits will always do more good than harm.

- *Utilization of tacit knowledge:* The self-organizing and self-governance of teams encourages them to find solutions for anything that might prevent them from meeting their iteration goals. As a result, the tacit knowledge of every team member can be utilized to solve the impediments faced by any of the team members.

- *User commitment:* The toughest thing about change is change itself, and what makes change really tough is lack of ownership. If the customer is not a part of the project, he will have no incentive to change willingly. With agile, the customer is part of the project team developing the product. The customer's feedback is sought and implemented throughout the project. The customer then feels some ownership of the project, and he thus has a vested interest in the project's success.

- *Knowledge management:* Agile does wonders for knowledge management. Its focus on communication and collaboration means that the team members communicate and learn from each other's experiences. Problems can be solved more easily and effectively when there is more than one brain working on them. This also results in maximum utilization of every team member's strengths.

- *The customer is indeed the king:* Agile methods literally put the money where the mouth is. A satisfied customer translates into ongoing and new business relations. And a satisfied customer is one whose needs and requirements are fully met by the product developed. With agile's focus on customer involvement, quick and frequent deliverables, and feedback into future development, the product developed is more likely to match what the customer envisioned it to be (and perhaps even more).

- *Improved team morale:* In more and more projects in the current environment, we are not dealing with unskilled laborers, but rather with knowledge workers. Knowledge workers are educated professionals who thrive on less hand-holding and more autonomy. Agile methods place significant value on the development team. They are allowed to work under minimal control, and their management trusts them to work optimally to deliver the required business value. This does wonders for the team morale and creates one of the biggest assets for any organization – a happy employee.

Chapter 2
Agile Methods

"If thy business be perplexed, divide it, and look upon all its parts and sides."

- Thomas Fuller, Comp., Introduction

Overview

When I talk of agile, I am immediately reminded of the orientation trips organized by many business schools these days. These trips take students to an outdoor spot, where they are given an adventurous assignment. Very few ground rules are set and then the team is left on its own. The team has to come together, know each other, utilize each ones strengths and then venture into unknown territory to achieve a known target. The team members communicate with each other and review performance all through the activity. The team can ask questions to the orientation masters during each activity, but no one is allowed to offer guidance or critique to the team during each activity. At the end of every activity, all teams come together, the orientation masters critique the performance of each team and they are set off to the next activity. Agile software development is quite like the above example. The project usually aims at developing innovative products, the teams are assigned what they are expected to achieve in each iteration, and the teams are self-organizing and work under minimal rules and oversight.

Principles of Agile Development

The principles of Agile development as per the Agile Manifesto (http://agilemanifesto.org), include: satisfying the customer, welcoming changing requirements, delivering working software frequently, engaging business managers with developers throughout the project, building projects around motivated individuals, face-to-face communications, working software as a measure of progress, sustainable development pace, continuous attention to technical excellence, simplicity, self-organizing teams, and regular reflection on how to be more effective.

The focus of Agile software development, as depicted in Figure 2.1, is defined by leaders of the Agile movement (at www.agilemanifesto.org) as:

Individuals & Interactions	>	Process & tools
Working Software	>	Comprehensive Documentation
Customer Collaboration	>	Contract Negotiation
Responding to change	>	Following a plan

Figure 2.1 – Focus of Agile software development – left over right, as per the Agile Movement leaders.

In other words, "while there is value in the items on the right, we value the items on the left more".

From the above, it is very clear that Agile software development aims at delivering innovative products to customers in a work environment that promotes communication, collaboration and adaptability.

Quite often, we have seen competition eclipsing co-operation. However, in the case of agile project management, competition and time-to-market pressure promotes co-operation and collaboration. There are a number of different approaches to agile project management. Some organizations use one of these approaches. Some use a combination of these approaches. Some tailor-make agile methods to suit their particular organizational needs. We shall now discuss some of the most well-known agile methods. Again, we must emphasize that while agile methods are more popular in the IT domain, the principles can be applied to other industries.

Extreme Programming (XP)

Extreme Programming is a discipline of software develop-ment based on values of simplicity, communication, feedback, and courage. It works by bringing the whole team together in the presence of simple practices, with enough feedback to enable the team to see where they are and to tune the practices to their unique situation.

-- Ron Jeffries

XP was developed by Ken Beck, Ron Jeffries and Ward Cunningham in the mid 90's while working on the Daimler Chrysler Comprehensive Compensation (C3) project. The highlights of XP are collaboration, simplicity and quick outcomes. There is nothing so extreme about XP except its focus on communication, communication and more communication. To understand XP better, let us go through the following thirteen XP practices, as illustrated in Figure 2.2:

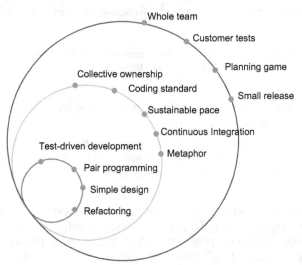

Figure 2.2 – The thirteen XP practices diagram – Adapted from www.XProgramming.com

Whole Team

One of the key highlights of XP is its insistence on on-site customer. A customer is not the one who funds the project but who uses the software. In XP, the team comprises of a customer, set of programmers, testers, and other such domain experts and/or managers as might be needed. The entire team is on-site. This arrangement facilitates communication and instantaneous feedback from the customer and among team members. This is in stark contrast with the waterfall methodology where customer feedback comes in only after the completion and implementation of the project.

Timely feedback helps in steering the project requirements such that they meet customer needs as closely as possible. Further, since the customer is available all the time, various ideas and alternatives can be discussed as they strike without having to wait till the next customer meeting.

Planning Game

In XP, planning is done in the form of short simple plans aimed at tracking what needs to be done in the near term to realize maximum business value. Planning is done dynamically, making use of the knowledge of the system developed so far and the changing business environment. Planning is done in the form of a game that happens once per iteration. A planning game consists of two parts:

- Release planning: The customer gives a list of desired features or functionalities. The developers give rough cost estimates to the customer for the functionalities to be developed. The customer determines what functionalities will deliver maximum ROI and prioritizes them for development.

- Iteration planning: The customer presents a list of required features for the current iteration in the form of story cards. Story cards are simple explanations of the functionalities to be developed. The team then does a detailed planning of tasks involved to develop the functionalities, as per the story cards, and delivers complete, running, tested features at the end of the iteration.

Small Releases

In XP, a project is divided into iterations that can be as small as one week long. At the end of every iteration, the development team is to deliver complete functionalities that render business value to the customer. Quick and small releases ensure quick feedback that can be incorporated into future iterations. They also keep the customer interested and committed to the project.

Customer Tests

When the customer gives a list of desired functionalities during the planning game, he also defines acceptance tests that the code delivered must pass for the feature to be acceptable. The development team then builds and runs those tests, just like the unit tests defined by the programmers themselves. This test further gives feedback to the development team about whether the team has

---

understood the customer's requirements and whether the feature developed delivers what the customer desires.

Test-Driven Development (TDD)

Test-driven development is one of the main aspects of XP, and reinforces its emphasis on quick feedback. In TDD, test cases are written first and then the software code, such that it passes the tests. Since the test cases are written even before the code, this works as a method of rapid feedback. Every build must pass every programmer's test cases. This heavy emphasis on testing ensures that the functionality delivered at the end of the iteration is robust and free of bugs.

Metaphor

XP teams develop a common vision of how the program works. They use a common system of names and descriptions that is understood by all the team. This system is similar to acronyms used by college kids for their friends and almost everything else. These acronyms are common knowledge to all in their group, and even a newcomer can quickly pick them up. The system metaphor is like a common story of how the system is expected to work. It is so easy to comprehend that any of the team members can guess the functionality of the class/method from its metaphor or common name alone.

Pair Programming

Pair programming is a software development practice in which two programmers work together on the same piece of code, sitting side-by-side on the same computer. Thus each line of the code gets reviewed even as it is being written. The code writer is called the driver and the reviewer is called the observer or navigator. The two programmers switch roles and also switch teams. This helps in dissipating knowledge fast. Every programmer has certain areas of expertise, and working in pairs enables a sharing of the expertise. Switching pairs enables further diffusion of expertise.

Pair programming also enables the driver to focus solely on the code while the observer works on reviewing the code for possible defects, and ensuring that the code matches the strategic direction of the project. He can also steer the driver back to the task at hand if the driver gets carried away by the code, and vice versa.

Pair programming has been found to deliver better code at no additional cost. However, as can be expected, not everyone is comfortable with the idea of having every keystroke watched and scrutinized. In addition, the pair must gel well and try not to be unnecessarily critical of each other's work.

Collective Ownership

In XP, all code is owned by everyone. Everyone is allowed to make changes to the code. This improves code quality, since every piece of code goes through several sets of eyes. If any bugs are found, the finder can correct it then and there, and not have to pass it back to any individual code owner. Pair programming complements collective ownership. Since programmers switch pairs, they get to see all parts of the code. Further, pair programming reduces the risk of altering code that is not fully understood.

Simple Design

Keep it simple is the mantra for Extreme Programming. Only as much code is written as is needed to deliver the required functionality. Duplication and unnecessary cluttering is avoided. There is no merit in unnecessarily complicating code when simple solutions can be just as effective. Further, a complex design makes future debugging, design changes and code modification difficult, and eventually makes the software archaic.

Continuous Integration

In XP, the code is kept fully integrated at all times. The greatest advantage of this is that there is less loss of valuable time and a decrease in frustration amongst programmers, as they are not hit by integration hiccups after several days of coding. Continuous integration also ensures that all possible bugs that can creep in during integration have been encountered well before final integration and delivery. As a result, there will be no last minute system breakup and subsequent embarrassment for the development team. Continuous integration further means that the programmers who build the code also integrate it with the rest of the system, rather than someone else who is not aware of the system. It is easier for these programmers to locate and correct bugs.

Refactoring

Keeping in mind the simple design motto, refactoring is an ongoing process in XP. Refactoring is essentially the process of cleaning up code to make it clear, concise and simple. Refactoring improves the source code without changing the end results of the functioning. Refactoring improves the comprehensiveness of the code, and cleans it up to make it free from dead code, inconsistencies and duplication. It makes the code easier to understand, maintain and enhance.

Coding Standard

In order for XP teams to collectively own the code and use a common set of metaphors, there has to be a standard for writing the code that is simple and easy for the entire team to understand and follow. A coding standard is a set of rules that the entire development team agrees to adhere to throughout the project. A coding standard improves the consistency and understandability of the code, and makes it seems as though it has been written by one single programmer. No particular coding standard needs to be imposed on the team; they can define their own standard or follow a pre-defined one. The most important thing is that everyone religiously conforms to the coding standard.

Sustainable Pace

A good development team is really one that can deliver good software again and again. A team that is overworked, stressed or tired will produce software that is buggy and of poor quality. The team may work overtime to meet budget and schedule goals, but the cost of repairing the damage done in the process, such as buggy software, an exhausted team and an unhappy customer, will far exceed the benefit. Thus, XP is a proponent of a steady and sustainable pace, one at which the team can repeat its success again and again.

Scrum

Scrum is one of the agile methods gaining a lot of popularity lately. It is an iterative development process, with each iteration resulting in a complete subset of the final product that can be implemented to deliver maximum business value early on. Scrum is discussed in greater detail in Chapter 3: "Scrum Demystified."

Open UP

The agile methods discussed so far seem too good to be true. And indeed this is the case for several organizations, who are just not ready culturally to switch to agile. The shift of focus from documentation to working software makes the management of some organizations very nervous. The above two processes place a lot of emphasis on tacit knowledge and its translation into explicit knowledge. This may render these methods less than effective for some projects, where there may be no significant tacit knowledge to start with, or it might be difficult to translate into explicit knowledge.

There is also the IBM Rational Unified Process (RUP), the pioneer in iterative software development processes. RUP has a guideline, or a process framework, for practices that covers a wide range of industries and projects, which a team can choose from for its specific project. However, its complexity and size have made it difficult to adopt for most small projects. OpenUP seems to bring some method to this madness.

OpenUP is rooted in the RUP. It adopts the best practices of agile methods and RUP to create a software development process that is lightweight and based on an open-source environment. In other words, OpenUP is an agile approach to RUP. The result is a win-win situation for all in the organization –management gets a stable, well-defined development process with just enough governance, and the software team works in an agile environment that improves performance.

Obviously, the biggest advantage of OpenUP is that it is based on an open source, enabling it to be used by all at no cost. Also, OpenUP enables the sharing of industry best practices, thereby enhancing OpenUP for future users.

OpenUP is a lean Unified Process that applies iterative and incremental approaches within a structured lifecycle. It places less emphasis on tools and ceremony, and hence can be extended to address a broad variety of project types. Essentially, OpenUp is an agile, lightweight process that encompasses well-known best practices in software development. It is based on iterative use case-driven development, short feedback loop and adaptability, team communication and collaboration, continuous testing and integration, and frequent delivery of complete functionalities. It provides guidelines on the fundamental needs of most projects.

However, it is not all-encompassing, in the sense that it does not cover all possible situations that may arise in a project. For such situations, OpenUP can be used as the base process, augmented by tailor-building other process content. The four core principles stipulated by OpenUP are:

- **Collaborate** to align interests and share understanding
- **Evolve** to continuously obtain feedback and improve
- **Balance** competing priorities to maximize stakeholder value
- **Focus** on articulating the architecture

All the above principles have direct correspondence with the Agile Manifesto.

Use cases, risk management and an architecture-centric approach form the basis of OpenUP. While architecture is given significant value in the OpenUP method, it is not treated as a large-scale frozen design of the system developed up-front, as in waterfall methods.

Instead, it just focuses on the principal requirements and constraints of the system that can be determined early on. This, like everything else, is an iterative and incremental activity, and how rigorously it is adhered to is determined by the complexity of the particular project.

The organization of work in OpenUP can be elaborated as below:

- **Project Lifecycle:** An OpenUP project is categorized into four phases: Inception, Elaboration, Construction, and Transition. The purpose of the Inception Phase is to define the scope and objectives of the project. A more detailed understanding of the system requirements occurs in the Elaboration Phase. Here an executable architecture for the system is created and use cases are defined. The iterative development of the complete system solution encompasses the Construction Phase. Finally, beta tests and deployment occur in the Transition Phase, and the system is thus released to end users. Each of the above phases consists of one or more iterations.

The project lifecycle provides a strategic understanding of the project and where it stands in regard to deadline. It also enables the stakeholders and team members to steer the project such that it delivers maximum value and encounters minimum risk.

This lifecycle model is where OpenUP differs from other agile methods. The lifecycle model steers early project efforts towards understanding the scope of the project and the solution envisioned before embarking on all-out development. This helps in mitigating maximum risk early on, even before real development work begins. As a result, real business value is realized soon after the construction iterations begin.

- **Iteration Lifecycle:** Just like in agile methodology, projects in OpenUP are completed in various iterations. Each iteration is a time-boxed period of a few weeks, during which the team develops a complete, potentially shippable subset of the final product. Each such subset is intended to deliver incremental business value systematically to the stakeholders. The team self-organizes itself in a way that it delivers the value that it has committed to. The iteration lifecycle determines the various micro-increments the individual team members shall work on to deliver the iteration objectives.

- **Micro-Increment:** Micro-Increment, as depicted in Figure 2.3, represents the personal contribution of team members to the OpenUP project. This essentially gives value to every effort, no matter how little, since it brings the team one step closer to achieving its target. This time frame for this result ranges from a few hours to a few days. It is also an effective short feedback loop that enables team members to adapt and enhance performance quickly. Team members share their daily progress on micro-increments, enabling collaboration and knowledge sharing.

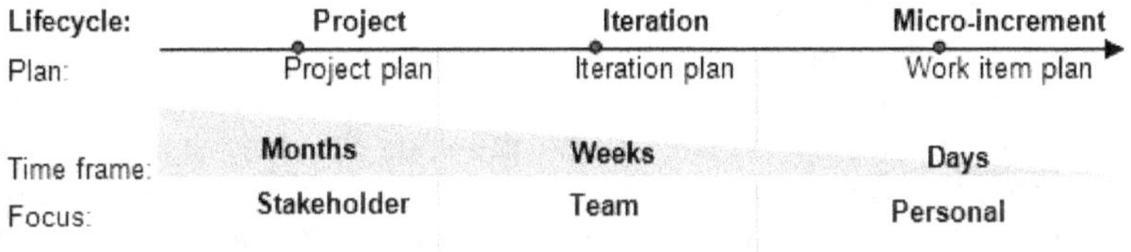

Lifecycle:	Project	Iteration	Micro-increment
Plan:	Project plan	Iteration plan	Work item plan
Time frame:	**Months**	**Weeks**	**Days**
Focus:	Stakeholder	Team	Personal

Figure 2.3 – OpenUP phases

Lean Programming

Lean software development is an adoption of the lean manufacturing practices of Japanese automakers Honda and Toyota to software engineering. Mary Poppendieck and Tom Poppendieck are the chief proponents of this method. Lean principles are not a software development method in themselves, but are successfully being applied to other agile methods. (Indeed, lean is not just for Hollywood actresses but also for software development.)

Lean software development aims at a shorter lead time of software delivery, higher quality of deliverable, and lower impact on budget through elimination of waste and reduction of defects in software development. Lean methods are based on the following seven principles:

Eliminate Waste

Lean thinking in manufacturing and software development classifies anything that does not add value (as perceived by the customer) to the product as a waste. Unnecessary, incomplete or defective code, or code developed before it is really needed, is a waste. The unnecessary handing-off of coding between teams is also considered a waste. Collecting requirements long before coding them is to begin is a waste. The same goes for managerial oversight in excess of what is required, as well as inefficient or insufficient communication. Any documentation over and above the minimum required is a waste. In other words, anything that does not deliver business value to the customer is a waste, and needs to be eliminated. The lean method requires that only such activity that adds value, and adds it in the best possible way, may be undertaken.

Amplify Learning

Software development is a continuous learning exercise, for both the developers and the customer. The developer learns from his and his team's experiences in the previous and ongoing iterations, and from the feedback he receives from the end users. The customer learns, and thus refines, his requirements from the output of the previous iterations. The learning process is expedited due to the existence of iterations, collaboration between the software team and customer, frequent testing, refactoring, and integration. There is no doubt that the amplified learning of all the stakeholders will improve product quality, resulting in higher business value.

Decide as Late as Possible

When developing an innovative product, it is likely that the product requirements will get revised quite often. Some of this is due to the knowledge gained during development iterations; however, most of it is due to a better understanding of the evolving market, environment or system itself, and an eventual tailoring of the initial requirements. Also, while the future is full of uncertainties, those uncertainties reduce as the future translates to the present. In such situations, delaying decisions until they are based on a degree of certainty will help reduce unnecessary iterations. This does not mean that decisions must be indefinitely delayed, since perfect requirement specifications will probably never occur. What this does means is that you should not jump to conclusions (coding) with a half-baked story, and repent later when the real story unfolds.

Deliver as Fast as Possible

A shorter cycle is the primary premise on which iterative software development is based. Short development cycles amplify learning by providing quicker feedback that can be incorporated into future iterations. In an environment where the highest market share and possible patents/licenses usually go to the organization that gets to the market first, companies cannot ignore length of delivery cycles. In fact, it would not be wrong to say that quick delivery has become the primary factor when it comes to selecting software vendors. It might be argued that quick deliveries could lead to a higher amount of bugs. However, with quick feedback, the magnitude of harm done is minimal and the bug can get truncated very early in the development process, as compared to the harm done if the bug goes unnoticed for a long time.

Empower the team

In order to achieve successful software development, organizations have begun to realize that developers can not be treated as replaceable commodities; rather, they are the basic nerve cell of the project. In order for projects to succeed, lean programming prescribes that development teams must be empowered to organize themselves, and take part in planning and design decisions. The principles of delayed decision-making and short delivery cycles do not go hand-in-hand with central monitoring and control; instead, they put a premium on autonomy and team integrity.

Build integrity

Perceived integrity is the customer's view of the system. If the customer feels that the system delivers what he wanted it to, he will perceive the system to have integrity. Conceptual integrity means that the system's central concepts work together as a smooth, cohesive whole, and it is a critical factor in creating perceived

---

integrity[2]. Software with integrity has a coherent architecture, scores high on usability and fitness for purpose, is maintainable, adaptable and extensible[3]. Effective communication is the most important prerequisite for a well-integrated system. There has to be a transparent flow of information between the customer and the software developers. The developer can then factor the feedback into software design and development. The developer also needs to regularly refactor and integrate the code to ensure that the whole works just as smoothly as the individual subsets.

See the whole

The whole can very well be greater than the sum of parts, since the whole is also comprised of the interaction of individual parts. However, for this to be true, each individual part has to keep his personal interests aside and do what is best for the whole. In a large complex project, there can be more than one team working; some teams can be from within the organization and some teams can be from sub-contracted organizations. However, all of the teams must work cohesively to produce the most optimum outcome for the organization. One of the very effective ways for achieving this is to base individual rewards on group outcomes.

Feature-Driven Development (FDD)

One of the principal concerns with most agile methods is scalability problems. Also, most agile methods are a tough sell to die-hard waterfall method followers. The lack of control and detailed planning makes them nervous that surprises might spring up at the most inconvenient moments.

FDD is a very palatable solution for such situations. FDD is an agile methodology introduced by Jeff De Luca and Peter Coad while working on a software project for a bank in Singapore in 1997.

FDD is a good solution for organizations that wish to go agile, but not all the way. They do like the idea of quick deliveries and customer involvement but do not wish to relinquish control of projects to development teams. They do want to be able to adapt to changing requirements but are uncomfortable with the lack of detailed requirement gathering and planning at the start of agile projects.

---

[2] Brooks, Mythical Man Month (1995) p 255
[3] Mary Poppendieck and Tom Poppendieck, Thinking Tools for Agile Software Development Leaders (2003)

Stephen Palmer and John Felsing give the following description of FDD in their book, A Practical Guide to Feature-Driven Development[4]:

> *"FDD starts with the creation of a domain object model in collaboration with Domain Experts. Using information from the modeling activity and from any other requirements that have taken place, the developers go on to create a feature list. Then a rough plan is drawn up and responsibilities are assigned. Now we are ready to take small groups of features through a design and build iteration that lasts no longer than two weeks for each group and is often much shorter—sometimes only a matter of hours—repeating this process until there are no more features".*

FDD sets aside a short time at the start of a project to allow the team and customer to share a common understanding of the domain in which work will be done. This results in the creation of an initial model and feature list. Thereafter, work begins iteratively on each feature. Figure 2.4 illustrates the FDD process in action.

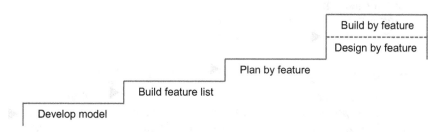

Figure 2.4 – FDD process in action

FDD does not require that all team members be highly seasoned programmers who can work well without being monitored. FDD can also be scaled up for use by large teams, or teams that are dispersed geographically, by breaking up a larger team into small feature-focused teams.

The following sections list best practices that are prescribed for organizations interested in following FDD:

---

[4] Prentice Hall PTR, ISBN: 978-0130676153

Domain Object Modeling

Domain object modeling consists of building class diagrams depicting the significant types of objects within a problem domain and the relationship between them[5]. The purpose of this practice is to construct a basic framework within which the system features are built. It is like creating a skeleton that gets fleshed out in later steps. The model helps all involved in the project to share a common understanding of the domain space and remove any misconceptions about what the customer's requirements are.

Developing by Feature

The term feature in FDD is very specific. A feature is a small, client-valued function expressed in the form:

<action><result><object>

with the appropriate prepositions between action, result, and object[6].

Features are essentially the decomposition of the domain model into small functionalities that the customer attributes business value to, and that can be completed in short iterations of two weeks or less. Feature is analogous to story cards in Scrum. It is something that the team can use to plan development and assign tasks, and the customer and management can use to track progress.

Individual Class (Code) Ownership

Readers might remember that most of the other agile methods propose collective ownership of all code. FDD presents a different view of this. One of FDD's principle practices is individual ownership of classes of code. The idea behind this is to resolve the issue of collective ownership dissolving into no ownership or ill-distributed ownership.

Individual ownership means that every piece of code can be attributed to a single programmer, who is responsible for maintaining the integrity of the code, taking care of enhancements and changes, and offering an understanding of the code. The problem with individual ownership is a possible loss of knowledge if the owner

---

[5] Stephen Palmer and John Felsing, A Practical Guide to Feature-Driven Development
[6] Stephen Palmer and John Felsing, A Practical Guide to Feature-Driven Development

programmer leaves the project and delays are caused due to interdependencies.

Feature Teams

Feature teams bring a sense of collectiveness to the FDD project. The object model lists a set of features that need to be developed. The development in FDD, as discussed earlier, occurs in features. The development team is then organized into feature teams, with each feature team responsible for pre-determined features.

Each feature team is headed by a Chief Programmer. Each Chief Programmer then forms his Feature Team, which consists of class owners of code that needs to be developed for the specific feature. Each class owner can be in more than one feature team, although putting them in too many teams can result in spreading them too thin. The Chief Programmer can also be part of another feature team as a class owner.

The teams are dynamically created as needed and disbanded as the purpose is achieved. For the smooth functioning of feature teams there has to be significant collaboration between the teams, such that work and class owners are optimally distributed, as depicted in Figure 2.5.

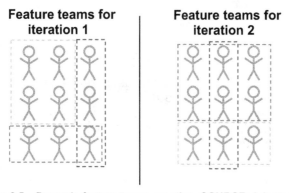

Feature teams for iteration 1

Feature teams for iteration 2

Figure 2.5 – Dynamic feature teams creation. SOURCE: Adapted from presentation on FDD by Justin-Josef Angel

Inspections

In FDD, inspections are carried out not just to detect defects and improve quality of code, but also as a tool to transfer knowledge and promote adherence to coding standards. It is important that inspections do not set off an alarm as performance reviews in the minds of developers. Inspections have to be viewed as collaborative debugging of code and a process for speeding up the learning curve.

Regular Builds

FDD prescribes that at regular intervals, all the developed features of the system be integrated together to build a complete system, albeit with limited functionalities. This helps identify integration problems early and also keeps the team ready to present the system to the customer, if required. An added advantage of this process is that with the developing by feature principle, the initial builds will consist of features that add high business value to the customer. Such builds, when demonstrated, will enable the customer to make more informed adaptations to the project, if required.

Configuration Management

While FDD requires only a history of source code to be maintained, this could be extended so that versions of all the various inputs that go into project execution are also maintained. Depending on the complexity, legal environment and agreement between the customer and the software vendor, various artifacts used in the project are version-controlled; in other words, maintained chronologically. These artifacts include requirements and design documents, test cases, customer contracts, etc.

Reporting/Visibility of Results

FDD prescribes that it is not just essential to know our goals. but also to know how we are doing in terms of progress towards achieving those goals. If a team cannot report its progress, it can be inferred that the team is listless about its performance and unable to control its activities.

FDD provides a simple low-overhead method of collecting accurate and reliable status information, and suggests a number of straightforward, intuitive formats for reporting progress to all roles, both within and outside of a project[7].

---

[7] Stephen Palmer and John Felsing, A Practical Guide to Feature-Driven Development

Dynamic System Development Method

Dynamic System Development Method (DSDM) is the extension, or rather, formalization of the Rapid Application Development (RAD) method that was implemented in England in the 1990s by the DSDM Consortium. DSDM is based on the following nine principles:

1. Active user involvement is imperative.

2. The team must be empowered to make decisions.

3. The focus is on frequent delivery of products.

4. Fitness for business purpose is the essential criterion for acceptance of deliverables.

5. Iterative and incremental development is necessary to converge on an accurate business solution.

6. All changes during development are reversible.

7. Requirements are baselined at a high level.

8. Testing is integrated throughout the lifecycle.

9. Collaboration and cooperation between all stakeholders is essential.

DSDM promotes the belief that software development is exploratory in nature and should not be expected to deliver perfect solutions in the first attempt. A DSDM project lifecycle is divided into three phases – pre-project, project lifecycle and post-project.

Pre-project

In the pre-project phase, project needs are identified, necessary approvals are obtained and resource commitment is ensured. As depicted in Figure 2.6, the pre-project work should be minimal, just enough to get the project off the ground and to ensure that all key stakeholders can be involved from the start of the Feasibility Study, thereby avoiding the need for later rework[8].

---

[8] DSDM Consortium Public Version 4.2

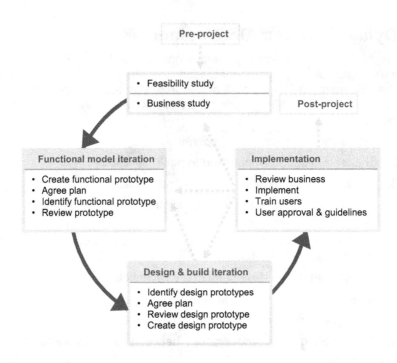

Figure 2.6 – DSDM lifecycle – Adapted from DSDM Consortium Public version 4.2

Project Lifecycle

The project lifecycle comprises five steps – feasibility study, business study, functional model iteration, design and build iteration, and implementation:

- **Feasibility study:** The basic purpose of the feasibility study is to assess if a technical solution to the business problem exists, and to give an initial estimate of resources, costs and timeframe required. In this step, it must also be decided if DSDM is the way to go for execution of this project. The reports generated during this stage include Feasibility Report, Feasibility Prototype, Outline Plan and Risk Log.

- **Business study:** This step takes the feasibility study from the previous step further by analyzing the characteristics of the business problem and the technical solution envisaged. Since this process is of short duration, workshops are organized to make the most of the short timeframe, wherein the technical and customer experts get together to collaboratively plan the system development. The reports generated during this stage are Business Area Definition, Prioritized Requirements List, System Architecture Definition and Development Plan.

- **Functional Model Iteration:** This is one of the first iteration steps in the DSDM lifecycle. The purpose of this step is to create a prototype of the system to be developed, based on the business study in the previous step. The prototype is built, demonstrated to users, and the feedback is incorporated. These steps are repeated till a mutually acceptable functional model is built. The end products of the iterations are the analysis model and various software components that satisfy some major functionality.

- **Design and Build Iteration:** This is the step where the actual development of the system happens. Some components that were built in the functional model iteration are refined here, while others for which only the analysis prototype was built in the previous step are coded here. The end result of this step is a tested system that can be delivered to the customer. The above two steps can be overlapping or happening concurrently, depending on the size and the needs of the project.

- **Implementation:** In this step, the developed system is put into actual use and the users are trained. The end-products of this step are the Delivered System, User Documentation and Increment Review Document. Based on the amount of outstanding work, if any, in the Increment Review Document, the project could either move to the post-project phase or go back to one of the steps in the project lifecycle.

Post-project

The purpose of this phase is to keep the system operating effectively by maintaining it. Maintenance can be done in a fashion similar to developing the system, using DSDM.

Crystal Clear

Crystal Clear is part of a family of agile software development methodology developed by Alistair Cockburn, based on ten years of research into successful projects. This methodology is applicable to small teams, comprised of 8 or less members, working on projects that are not life-critical. In his book *Crystal Clear: A Human-Powered*

Methodology for Small Teams[9], Alistair Cockburn describes Crystal Clear as:

> *"The lead designer and two to seven other developers ... in a large room or adjacent rooms, ... using such as whiteboards and flip charts, ... having easy access to expert users, ... distractions kept away, deliver running, tested, usable code to the users ... every month or two (quarterly at worst), ... reflecting and adjusting their working conventions periodically".*

The Crystal Clear method of software development lays significant emphasis on people and communication. Processes and artifacts are expected to support people, and can thus be molded as needed. Crystal Clear is based on the following seven properties. The first three are mandatory, while the next four are recommended.

Frequent Delivery

Crystal Clear, like all other agile methods, emphasizes frequent delivery, with the maximum period between deliveries being three months. Frequently delivered complete functionalities reduce the length of the feedback loop and keep all the stakeholders interested in the project.

Reflective Improvement

This principle complements the previous one in that it requires the teams to get together every so often to reflect on their performance, based on the feedback received and their personal experiences. The idea is to continuously improve performance even as work goes on for the project.

Osmotic Communication

This principle is meant to take benefit of the co-location of team members in the same room or, at worse, adjacent offices. By being physically in the same room, team members can learn from the conversations of others and possibly even solve their problems without even asking for help. This principle allows team members to communicate even while not directly communicating.

---

[9] Addison-Wesley (January 1, 2005)

Personal Safety

This is the first of the recommended principles. This principle requires that for the team members to really take ownership of the project, they need to be able to communicate openly without having to worry about the consequences. This means that they can even communicate inadequacies without any inhibitions.

Focus

While it is important that team members know the list of priorities, it is also important that they have the flexibility of time without external distractions, in order to be able to work on the priorities. A focused team will deliver not just quick but also quality results.

Easy Access to Expert Users

Feedback from real users of the system is very helpful in improving the final usability of the product. For getting such feedback, the users who really know the ins and outs of the system must be accessible.

Technical Environment

The three technical tools that this property talks about are automated tests, configuration management and frequent integration. All these tools are aimed at containing the pain of debugging way down the line by discovering bugs at regular, frequent intervals.

Chapter 3
Scrum Demystified

Scrum is an agile method for project management developed by Ken Schwaber.

What is Scrum

Scrum is an agile software development methodology that develops software in an iterative and incremental fashion. It is an empirical method and works very well in developing innovative products when product requirements are not clearly stated upfront. It gives a lot of authority to the team to manage its own work and prescribes only a very simple set of rules for the team to follow.

Scrum is an effective method in an organization that works on short projects or can break down long complicated projects into incremental, manageable modules. This reduces the time-to-market and the feedback loop.

Scrum is built on 30-day cycles, or "sprints," that deliver working increments of the final product. Thus the next increment is built based on requirement specifications as well as modifications to the specs resulting from what was learned in the previous sprint.

The Scrum Process

Figure 3.1 is a graphic representation of the entire scrum process. At the start of the scrum project, the product owner defines the current requirements of the project in the form of a list called Product Backlog. He prioritizes the list based on which functionality delivers maximum value. Scrum is based on the 80-20 principle. Quite often, 20% of the project functionalities can deliver 80% of the project ROI. In traditional project management, with 'big bang' implementation at the end, there is no way to know this or to benefit from it. However, with Scrum, since the product owner is involved in the planning of each sprint, the sequence of each sprint can be designed in such a way as to deliver maximum value early on in the project. In other words, "a bird in hand is better than two in the bush."

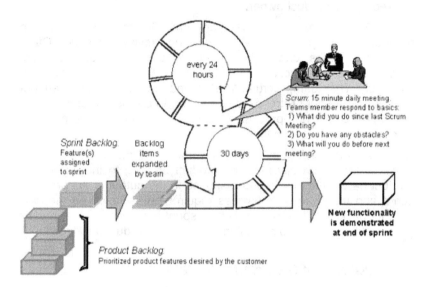

Figure 3.1 – The Scrum process - Source: www.controlchaos.com

The Scrum team then decides how many functionalities or features it can expect to complete in the first iteration, keeping in mind the priorities of the product owner. Each iteration is called a Sprint and is 30 consecutive calendar days long. In other words, each sprint is like a 30-day stop-watch. This list is then expanded to create a list of tasks needed to be performed in that sprint. This is called the Sprint Backlog.

The team is then left on its own to work on the sprint backlog. No outside interference or influence is allowed to affect the team's performance during the sprint.

During the sprint, the team gets together at the start of each day in a short meeting called the Daily Scrum. The purpose of the daily scrum is to summarize what was done yesterday, what will be done today and what problems need resolution to achieve the sprint targets.

The Scrum Master is the modified project manager for the Scrum project. He facilitates the team, makes sure that nothing can impede team performance and ensures that Scrum practices are adhered to. At the end of every Sprint, the team is expected to deliver a complete component of the final product that can be implemented if desired by the product owner.

After every sprint, the team, the Scrum Master and the product owner get together for a sprint review meeting. Other interested parties can also attend. In this meeting, the team demonstrates the functionality completed in that sprint and takes in feedback. The team, Scrum Master and product owner then get together for a sprint retrospective meeting to discuss what went well in the previous sprint, process-wise, and what needs to be modified to improve team performance.

At the next sprint planning meeting, the product owner presents the potentially modified product backlog to the team. The product backlog is modified in response to the output of the previous sprint and the changing business requirements. Based on the new product backlog, the next iteration or sprint is planned. This process goes on till there are no more items left on the product backlog.

Key Artifacts of the Scrum Process

The following are the key artifacts of the scrum process:

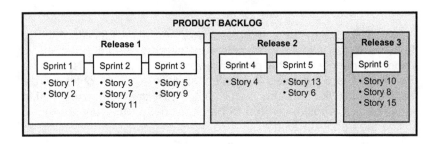

Figure 3.2 – Scrum iterations

Product Backlog

The product backlog is a list of product requirements. It is never complete and evolves for as long as the project is active. It is prepared and maintained by the product owner. The product owner prepares the initial product backlog that is used to kick off the first sprint. Thereafter, the product owner rearranges the product backlog, adding or deleting items from the list based on functionalities developed so far and on changing business needs.

Release

The product owner arranges the product backlog into a number of release cycles. The earlier releases are expected to generate higher ROI. Each release is comprised of a few sprints. A value-generating feature or functionality is delivered to the customer at each release.

Sprint

A sprint is like a stopwatch that stops at the end of 30 consecutive calendar days. A complete project can be comprised of many such sprints or iterations. At the end of each sprint, a fully tested and working subset of the project is delivered to the product owner, who can choose to have it implemented. The team chooses how many items or story cards from the product backlog the team will work on during any particular sprint. If, at the end of that sprint, there are items that were chosen to be worked on but could not be completed, they are returned to the product backlog and are re-arranged as required by the product owner. In the same way, if the team realizes, during a sprint, that it can deliver more than the functionalities promised, it can go back to the product owner and pick more story cards from the product backlog to be worked on in that sprint.

Story Cards

The items on the product backlog list are called story cards. They are a further explanation of the functionalities to be delivered. Story cards are arranged based on their value generation capability. In other words, stories expected to deliver maximum ROI are worked on with higher priority. Stories are started and completed within one sprint. If, for any reason, they cannot be completed in a particular sprint, they are returned to the product backlog and are worked on in future sprints.

Sprint Backlog

Once the team decides on which functionalities it will work on in a particular sprint, it breaks down the functionality into actual tasks that can then be worked on by the individual team members. The sprint backlog is also an evolving list and it evolves based on the team performance during the sprint. The team can add or remove tasks in the sprint backlog, rearrange them or reassign them.

Daily Scrum

The development team and the Scrum Master meet at the same time every morning for a short 15-minute review meeting called the Daily Scrum. The daily scrum keeps the team on track and helps them to evaluate performance and/or switch gears, if necessary, to meet the sprint goals.

The Daily Scrum meetings remind me of discussing my day at school with my mom as a child. My mom would always spend a few minutes with me when I got home from school each day. She would want to know what I had done in school that day, if everything went well, what were my tasks and schedule for the next day, and if anything was required to prepare me. These are quite similar to the questions asked by the Scrum Master to the team at the Daily Scrum:

- What have you done on this project since the last Daily Scrum meeting?
- What do you plan on doing on this project between now and the next Daily Scrum meeting?
- What impediments stand in the way of you meeting your commitments to this sprint and this project?

No digressions are allowed beyond these three questions. If issues or problems arise that need team collaboration, meetings between interested parties are arranged for a later time. The 'chickens' (see pg. 49) are allowed to be silent observers of the Daily Scrum, but they are expected to be just that – silent observers. They are not allowed to make observations or recommendations to the team during and after the Daily Scrum, or during the sprint.

Burndown Chart

A plot of the hours of work remaining in the sprint backlog against time is known as a Burndown Chart. A burndown chart is an excellent tool for analyzing the progress of the team in achieving the sprint goals. The point at which the chart meets the X-axis is the most probable expectation of the work completion date. A breakdown chart looks rather basic, but can be put to very good use in detailed analysis of team performance and work management. It can be used to do a what-if analysis to see how the chart and the intersection point change if functionalities are added or removed from the sprint.

The slope of the burndown chart, as depicted in Figure 3.3, is also very informative. It is inevitable that the chart line sometimes trends upwards. If, however, it does that quite often, it might imply that sprint planning needs to be done more rigorously. If the trend does not slope downwards smoothly and evenly, it could imply that the work done is not reported correctly, or that it is done in fits and starts. This goes against the principle of sustainable pace of development in agile methods. This, again, needs to be corrected.

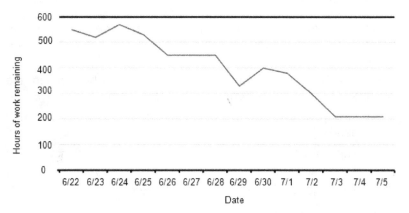

Figure 3.3 – The slope of burndown chart.

The Roles on Scrum Teams

Generally, scrum teams do not include any of the traditional roles for a project. For instance, they do not include, for software engineering, roles such as programmer, designer, tester or architect. Everyone on the project works together to complete the set of work they have collectively committed to complete within a sprint. Scrum teams develop a deep form of camaraderie and a feeling that "we're all in this together." A typical Scrum team is 6-10 people, but scrum professionals such as Jeff Sutherland have scaled Scrum up to over 500 people.

The Scrum participants can be divided into two groups, the pigs and the chickens, as Ken Schwaber calls them:

1.   The 'pigs' are the involved participants who really get their hands dirty.

2.   The 'chickens' have interests in the project but are not directly involved.

The 'pigs' are further divided into three roles:

- the Product Owner
- the Team
- the Scrum Master

The Product Owner

The product owner represents the various stakeholders of the project. He puts together the initial requirements of the project and procures the necessary funding. The initial project requirements are called the Product Backlog. The product owner frequently revisits the product backlog to make sure that the priorities are set correctly in the face of changing business requirements, and in light of the functional and non-functional components delivered so far. The product owner can request implementation at the end of any sprint, well before project completion. The product owner will exercise this option if doing so will lead to immediate realization of business value, and thus keep the 'chickens' hooked on to the project.

The Team

Scrum teams are cross-functional teams comprised of highly competent individuals capable of delivering high-quality innovative products. Discipline and integrity are key characteristics of a successful scrum team. The scrum team members work very well together in absence of outside control or management. The team is collectively responsible for successfully delivering a complete component of the project at the end of each iteration.

Scrum works best where the management is willing to relinquish control of the project to the project team. In a setting where there are many complex tasks being worked on at the same time, central control is inefficient. If a project is to succeed in such a setting, a sense of ownership needs to be instilled in the project team. And for this to happen, the project team needs to be self-organizing and self-controlling.

The Scrum Master

The Scrum Master is the proponent of the scrum process for the project. He ensures that scrum is understood and followed by the team members and the product owner. He shields the team from any outside interference and makes sure that the stage is set for successful completion of each iteration and the entire project.

Scrum in Non-IT context

Although applied commonly to IT projects, scrum is equally useful in non-IT projects. Scrum is more about team dynamics and work culture than about any specific project management practice. Scrum is a style of managing projects, and hence its principles can be applied in part or fully in any context. Scrum can be applied whenever the project can be broken down into meaningful increments. Take the simple case of home improvement projects that all of us indulge in at some point. These projects can be broken down into various activities, such as corking the holes, painting the walls, hanging the pictures, (re)arranging the furniture, etc. Each of the tasks is an item on the product backlog and each of the backlog tasks can be broken down into a list of tasks for the sprint. Each sprint makes a meaningful change to the house.

Scrum can be extremely useful in start-ups. The iterations effect quick feedback loops that can be imported to future tasks. This enables early track change or adaptations as required, and hence prevents unnecessary drainage of resources. This is very helpful for start-ups that have limited resources as it is.

Scrum can be applied to market research projects that precede new product launch and also to the entire new product launch project. For companies in the Consumer Packaged Goods

(CPG) industry, time-to-market is extremely important. Scrum fits the bill perfectly. Shorter iterations mean that the project team must be nimble enough to make rapid adjustments to meet market demand and compete effectively.

Some of the scrum principles can be applied to heavy engineering projects. These projects can be divided into tasks that form the product backlog, and these tasks must be prioritized. The priorities, however, might have to be defined more by what part of the machinery or product needs to be built before work can begin on the next, and less by what adds value to the product owner. The tasks on the product backlog get divided into sprints, and each sprint comprises of a number of sprint backlog items. There is a daily scrum meeting every morning. Each sprint delivery may not be implementable in the real sense but will definitely add meaningful value to the final product. It can be reviewed by the product owner and necessary changes may be made early on instead of allowing an oversight to trickle into the final deliverable.

We have discussed only a few applications of scrum in the non-IT context. However, as mentioned above, scrum or adaptations from it can be applied in almost any project.

Chapter 4
Conventional versus Agile Project Management

Hey! You guys lied! This isn't agile.[10].

Overview

A question that is always asked by organization leadership investing or considering investing in project management is, "Where is the value of such implementation to the project itself?" It is quite common to find operations professionals not willing to invest in project management, as they consider it an impediment to effectiveness. Often these professionals will argue that project management best practices, as per the Project Management Institute's (PMI) guideline, require too much time spent with paperwork, record keeping and analysis. Just as often, the project managers committed to follow PMIs Project Management Body Of Knowledge (PMBOK) are viewed as administrators who put together too much detailed task schedules, create colorful resource profiles, chase and disturb team members with micro-task completion information, and spend too much time writing volumes of status reports for sponsors and upper management. By the same token, project teams often view the investment in project management, including time, best practices, tools and other resources, as overhead.

---

[10] Wendy Friedlander, from her Wunda's World blog at http://wundasworld.blogspot.com/2008/11/when-agile-works.html

Rather than overhead, project management should be seen as a key component for inspirational leadership determined to deliver value to its customers. Why is it that so many resist project management then? In our view, the most likely reason is the fact that many project managers, and their practices, are focused on compliance activities instead of the delivery of value to their stakeholders. Indeed these professionals are often closer to project administrators than to project managers. Customers always pay for value, and view everything else as overhead, whether it is necessary or not. Sure, it's not difficult to explain the time and effort spent in activities or deliverables that help comply with government regulations, on documentation that conforms to legal requirements, but to the customer's point of view, none of it adds value. The same goes for status reports to assist managers in meeting their fiduciary responsibilities: as far as the customers are concerned, these reports don't add value. And the list goes on.

But how do we measure customer value, and project management's contribution to that value? Is the goal of project management best practices to have a project finish on time and under budget, while receiving the stamp of approval from its stakeholders and sponsors? Let's take the example of Motorola's disastrous, multibillion-dollar Iridium project, which was a major failure in the market. The project was actually considered a success by many, since it had fulfilled its original project scope. Meanwhile the movie "Titanic," which was severely over budget and extremely late, and viewed by many as a $200 million flop, was actually the first movie to generate over $1 billion in worldwide revenue. And so by common compliance-based project management practices of budget, scope, and schedule performance, "Titanic" was a failure. Was it?

The reality is that all projects face similar demands, varying from customer needs and profit to development speed and scope creep. These often contradictory constraints and demands on project execution — speed and quality, great functionality and low cost, uncertainty and predictability, mobility and stability — have created the need for project managers and project management practices that focus on delivering value.

It is this main premise, *a need for project managers and project management practices that focus on delivering value,* that agile project management (APM) aims to attain. As introduced in earlier chapters, APM as a set of values, principles, and practices can assist project teams in coming to grips with this challenging premise. The core values of APM address both the need to build agile, adaptable projects and the need to create agile, adaptable development teams.

Is Agile Enough?

For agile project management, agile development methodologies such as Scrum and eXtreme Programming (XP), as discussed earlier, tend not to be enough. Scrum is excellent for managing a project team's workload and delivering products incrementally through iterative development, while XP is excellent for agile engineering practices that improve product quality

Although PMI's PMBOK really embodies all the *traditional project management* best practices, and is a very useful resource, it is not enough for APM. No doubt it includes traditional project management practices that are not appropriate if you are doing agile, but it also includes key aspects of a project that need managing which are simply not addressed by Scrum or eXtreme Programming. These include, but are not limited to:

- Project Initiation
- Cost Management
- Human Resources Management
- Communications Management
- Risk Management
- Procurement Management
- Stakeholder Management
- Organizational factors

Sure, at this stage in the book, we all agree that in APM we don't want to see a big specification upfront, as the PMBOK would have it. We also don't want to see every task mapped out on a huge Gantt chart, nor do we want to see change control as the process for scope management. But we do need all the above areas managed in many agile projects.

So how do we actually plan for APM? At MGCG, our consulting practice overcame this by adopting *traditional* project management, as per the PMBOK's guidelines (or the PRINCE2 project management methodology that has become the standard in the UK), applying the relevant aspects of the traditional PM approach with the agile practices of Scrum and XP. We effectively augment agile with traditional project management methods where appropriate. This is to say that there is no *APM for Dummies*. Also keep in mind that APM isn't a silver bullet, and does not hold the answer to all your project challenges. At this stage, more than half ay into the book, we can summarize APM and the goal of its planning in two statements:

APM is the ability to both create and respond to change in order to profit in a turbulent business environment.

APM is the ability to balance flexibility and stability within project constraints.

An example of a product development effort in which all the aspects of agility come into play is that of small, portable DNA analyzers. These instruments, which are still five or more years away from commercial deployment, could be used for analyzing suspected bioterrorism agents (e.g., anthrax), making a quick medical diagnosis, or performing environmental bacterial analysis.

These instruments must be accurate, easy to use, and reliable under wide-ranging conditions, and their development depends on breakthroughs in nanotechnology, genome research, and micro-fluidics. Half a dozen companies are racing to build versions of a DNA analyzer, often with grants from agencies such as the US National Institute of Health. Developing these leading-edge products requires a blending of flexibility and structure, exploration into various new technologies, and creating change for competitors by reducing delivery time. These are not projects that can be managed by traditional, prescriptive project management methodologies.

Some people mistakenly assume that APM connotes a lack of structure; hopefully by now you are not one of them, since the absence of structure, or stability, generates chaos. Conversely, too much structure generates rigidity (traditional PM methods). Complexity theory tells us that innovation — creating something new in ways that can't fully anticipate an emergent result — occurs most readily at the balance point between chaos and order, between flexibility and stability. Scientists believe that emergence, the creation of novelty from agent interaction, happens most readily at this "edge of chaos." The idea of enough structure, but not too much, drives APM managers to ask the question, "How little structure can I get away with?" Too much structure stifles creativity; too little breeds inefficiency.

This need to balance at the edge of chaos to foster innovation is one reason process-centric methodologies often fail. They push project managers into over-optimization at the expense of innovation. APM users do not get lost in some gray middle ground; they understand which factors require stabilization and which ones encourage exploration.

For example, in a high-change product development environment, rigorous configuration management stabilizes and facilitates flexibility just as a focus on technical excellence stabilizes the development effort. The concepts and practices described in this book are designed to help project teams understand this balancing between flexibility and stability. They help answer the question of what to keep stable and what to let vary.

APM therefore means being responsive or flexible within a framework or context. As we have discussed in earlier chapters, the problem with many traditional project management approaches is that they too narrowly define the context; that is, they've planned projects to a great level of task detail, leaving very little room for APM to be effective.

Balancing at the edge of chaos, between flexibility and stability, requires project managers who are great at improvising, who have the ability to deal effectively with the ambiguity, and the paradox, of pursuing two seemingly dissimilar goals at once. Project organizations that support these improvisers have three key traits:

- An adaptive culture that embraces change

- Minimal rules that encourage self-organization, combined with the self-discipline to closely adhere to those rules

- Intense collaboration and interaction among the project community

The APM Planning Framework

Project management processes and performance measures are different for exploration- and experimentation-based approaches than they are for production- and specification-based ones. Production-oriented project management processes and practices emphasize complete early planning and requirements specification with minimal ongoing change. Exploration-based processes emphasize nominal early planning, good enough requirements, and experimental design with significant ongoing learning and change. Each approach has its place, but the lifecycle framework for the latter has a very different flavor from the former. The APM framework consists of five phases, each with supporting practices. They are Envision, Speculate, Explore, Adapt, and Close.

These phases resemble a scientific investigative process more than a production management process. The Envision phase results in a well-articulated business or product vision, enough to keep the next phases bounded. In the Speculate phase, the team hypothesizes about the specifications of the product, knowing that as the project continues, both technology and customer specifications will evolve as new knowledge is gained.

The Explore phase then becomes a parallel and iterative operation in which the preliminary specifications and design are implemented. Components labeled "uncertain" are subject to more experimentation than others whose specifications or designs are more certain. In the Adapt phase, the results of these experiments are subjected to technical, customer and business case review, and adaptive actions are incorporated into the next iteration.

One of the most common questions about APM is, "What about the planning, architecture and requirements phases?" The simple answer is that these things are activities and not phases. An agile approach can easily include as much time for these activities as in a conventional serial phase approach, but the activities are spread across multiple iterations.

A second area of concern is the risk of rework in agile development if the initial architecture work (the discussion in this section could refer to architecture, plans or requirements) misses a critical item. But there is an equal if not greater risk in serial development that often goes unnamed — that of getting the upfront architecture wrong. In a serial process, validation of the early architecture decisions comes late in the project lifecycle, when the actual building occurs. By then a tremendous amount of time and money has been spent. Changing the architecture then becomes a major, and costly, decision — if it is possible at all.

Within this general APM framework, the successful completion of each phase depends upon a series of key practices that actually guide the work effort. Other practices, such as customer focus groups and iterative planning, are based directly on the core values of agile organizations, while such practices as workload self-management and participatory decision-making focus on building an agile, adaptive culture. Values and guiding principles describe the why of APM, and practices describe the how.

The Conventional Way: Optimization versus Adaptation

Figure 4.1 depicts conventional project lifecycle stages. As the figure illustrates, any project goes through these four distinct stages throughout its lifecycle. In each one of those stages a series of tasks are implemented, and with that the level of effort invariably tends to increase. A well-managed project lifecycle will always display a bell curve, as shown in picture 4.1, and it will always strive to optimize that bell curve.

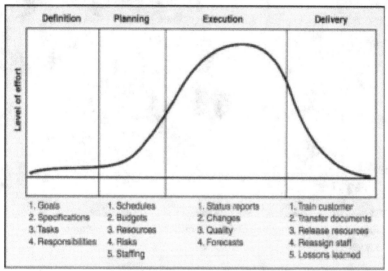

Figure 4.1 – Characteristics of a Conventional Project Management

Of course, this will only happen in an ideal world; still, the closer you can get your project lifecycle to display a bell curve, the more smooth and well-balanced your project execution will be. It is here that the CMMI model will work well, as organizations and project managers strive to climb up the levels of the model all the way to the fifth level: continuous improvement of process through optimization.

Figure 4.2 provides a summary of the main characteristics of conventional project management, in our view:

- Project life cycles comprises "seven" phases.

- They are built for known functionality. Eventually there is a time for "ready."
- The process focus is more on optimization than adaptation.
- Change is more of an exception than the norm.
- There is a tendency for a predictable future (lifetime).
- Matured technologies.

Conventional PM

- Made of "seven" phases..Requirements, Design, Operations
- Build for known functionality..There is time for "ready"
- Process focus is more on optimization than adaptation
- Change is more of an exception than the norm
- Predictable future (lifetime)
- Matured technologies

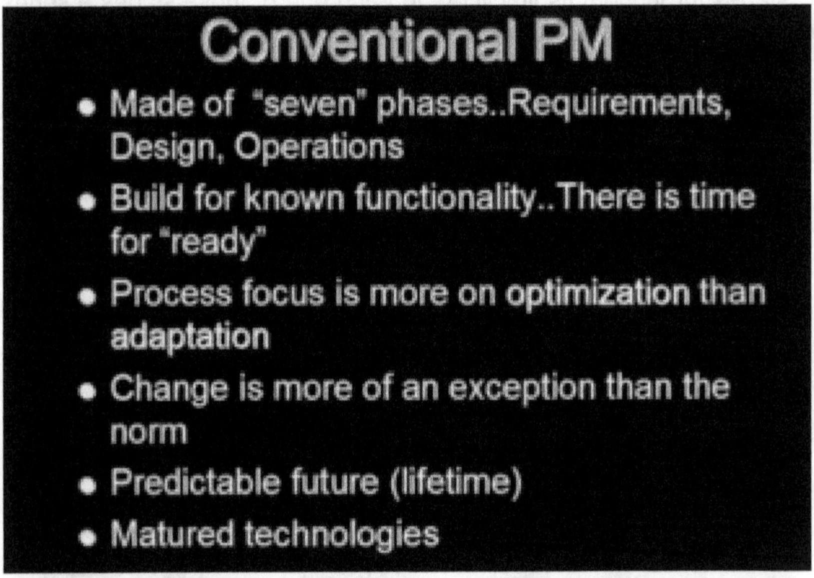

Figure 4.2 –Characteristics of a Conventional Project Management Approach

Let's review each one of these characteristics before we take a look at APM differences.

Project Life Cycle Comprised of Seven Phases

Although we acknowledge the technical correctness of the PMI's PMBOK in defining the project lifecycle as having four major stages (Definition, Planning, Execution and Delivery), in our experience we consider "seven" (notice the quotes), the four authentic and tangible stages defined by PMI and three more pseudo-phases, so to speak.

To clarify, let's do an overview of PMBOK's four phases and see where we introduce the other three:

1. **Definition:** The first phase, where all the specifications of the project are defined, project objectives are established, teams are formed and major responsibilities are assigned.

2. **Knowledge Management (KM):** To me, the second "best-practice" phase. Here we build the necessary KM systems, as discussed in the previous section, establish mentorship relationships and apply our Knowledge Tornado methodology in a attempt to turn every PM member's knowledge into action, trying to capture not only their explicit knowledge but tacit as well.

3. **Planning:** The official second phase (my third), where plans begin to be developed, the level of effort (LOE) increases, and the goal is to determine what the project entails, what its schedule looks like, who are the stakeholders, who does it benefit, what will be the quality level to be maintained, and what is the budget.

4. **Change Management:** To me the fourth stage, not explicitly defined on PMBOK four-phases, but loosely embedded throughout the phases. Here project staff, environment, stakeholders, sponsors and the project itself is assessed to determine the degree of change that will be promoted through the execution and successful delivery of the project, and what proactive actions must be taken in order to increase the level of success of the project as a whole, and prevent resistance to change and/or new benefits introduced by the project.

For example, we were once retained on a huge user application migration that did not take into consideration change management. Technically the project was a success, but as far as the bottom-line was concerned the project was a major failure, and required a new project to mitigate all the problems, mostly resistance to change, created by the original project.

This all happened when Fleet Bank acquired Bank of New England (BNE) in Boston and decided to migrate all BNE users from MultiMate word processing to WordPerfect. Technically the project was well thought out, and in a single weekend we migrated all users onto the new WordPerfect platform, converted their files, and installed new templates to ease their initial adaptation to the new system. Half-day training was also provided on the use of WordPerfect.

The problem was that no one bothered to assess what the reaction of users would be when asked to effectively change the way they worked and utilize a whole new system they had no idea about. There we also several different groups of users, some clearly needing more than half-day training and some who barely needed training at all.

In addition, users could not understand why the changes needed to take place; some felt they were being pushed out of the organization, others felt their performance would suffer tremendously, causing them to miss their deadlines, and other were just plain lukewarm about the whole transition, as they felt these changes were being imposed on them.

As the popular saying goes, "whatever is imposed is also opposed." As a result, very few users were actually using the system in a functional manner; many were wasting unbelievable amounts of time trying to recreate the familiar environment they knew in MultiMate, while some refused to work on the new system at all, choosing to continue to work on the old system underground and submit converted files to peers when necessary. Fleet Bank was therefore forced to implement change management techniques, varying from just listening to frustrated users and pledging full support during their transition, to extended training broken down into three different categories, and in some cases, believe it or not, allowing the co-existence of both systems.

5. **Execution:** Actually the fourth level of the project lifecycle, this is where the major portion of the project work takes place. Time, cost and specification measures are used for project control and performance assessments.

6. **Risk Management:** This is my sixth stage, an embedded process under PMBOK's model. Although risk management is a necessary step during project definition/planning, in reality there are several risks that are near to impossible to detect, never mind have a mitigation plan for, prior to the actual start of project execution. Therefore, I always recommend keeping a close eye on risk management throughout the project and having the well-know risk matrix converted into a dynamic system, much like the Department of Defense's DefCon System. There will be days when the whole project is operating under green alert, other times under orange, but hopefully never under red alert.

7. **Delivery:** This is my seventh and final stage, the fourth under PMBOK, which typically includes two main activities: the delivery of the project and the redeployment of project resources. In our practices, knowledge and change management play a major role here again, in the process of knowledge transfer and in making

sure that customers and stakeholders can fully benefit from the project and the changes it introduces.

In practice, the project lifecycle can not only be broken down into four to seven stages, it can also be used by many project groups to depict the timing of major tasks over the life of the project.

Built for Known Functionality

Typically, conventional project management is built and developed with a know functionality, which is usually the result of a vision and the establishment of very tangible objectives. Objectives translate the organization mission into specific, concrete measurable terms.

Therefore, the whole project definition works around the understanding of the vision and the ability to translate it into an attainable objective. This objective becomes the goal, the focus of the project, setting targets for all levels of the organization. These objectives also pinpoint the direction project managers believe the project should move toward.

In order to be successful, the project manager needs to make sure those objective are achieved in a timely fashion and within the budget set during the definition stage. To achieve a project objective means to enable certain functionalities, the benefits of the project, to be obtained. A project will be successful if it meets the goals – list of functionalities - set during the definition stage.

In summary, in a conventional project, the functionalities are known from the start through the objectives, which will determine where the project is heading and when it is going to get there; a baseline is established. Any variation from that baseline causes a series of reactive and corrective actions, so that the objective can still be achieved.

Therefore, when setting objectives, make sure their characteristics are SMART (Specific, Measurable, Assignable, Realistic, and Time-related):

- **Specific** – Be specific in setting your objectives.

- **Measurable** – Establish a measurable indicator(s) of success.

- **Assignable** – Make sure the objective is assignable to a person for completion.

- **Realistic** – State what can realistically be done with available resources.

- **Time-related** – State when the objective can be achieved, its duration.

Focus on Optimization Rather than Adaptation

A conventional project management approach focuses on optimization rather than adaptation. Strategies are formulated to include well-determined and evaluated alternatives that support the project's objectives and ensure the best alternatives are selected.

Cause-and-effect worksheet diagrams, as shown in figure 4.3, are often used to develop a priority system that helps optimize project execution and troubleshoot problems along the way. The fishbone diagram is also an excellent tool for analyzing and isolating symptoms and causes of problems.

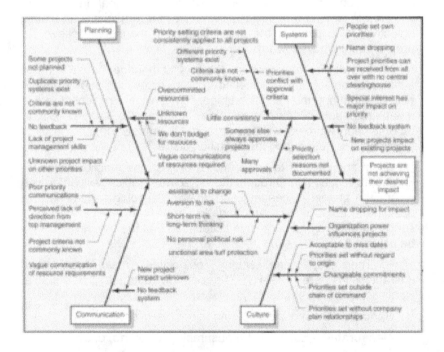

Figure 4.3 - The cause-and-effect diagram can help develop priority system to optimize project execution

Change is More an Exception than the Norm

In a conventional project management system change is more an exception than the norm. Thus a myriad of evaluation and control mechanisms play an active role in any project, requiring a single information system that measures project progress and performance against a project plan that supports the delivery of the project on time, on budget, and in the form requested by the client.

Project schedule control charts, such as the one in figure 4.4, are one of many tools used to monitor project schedule performance against current performance, and to estimate future schedule trends.

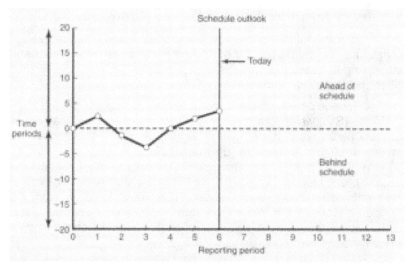

Figure 4.4 – Project schedule control charts are used to plot the difference between scheduled times on the critical path at the report date with the actual point on the critical path.

In an attempt to prevent any changes in the expected project tasks and deliverables, systems monitoring project performance attempt to periodically provide project managers and stakeholders with:

- The current status of the project in terms of cost and schedule
- The forecasted cost of the project at completion time

- The estimated date of project completion
- The exposure of, and alert to, any potential problems that should be addressed immediately
- The reasons and locations of any potential cost and/or schedule overruns
- The return on investment (ROI) for every dollar spent in the project
- Whether or not potential problems can be identified before they strike, or it becomes too late to correct them.

There are several simulation software packages available on the market so that conventional project managers can establish *what if* scenarios; SimProject is one of them.

Tendency for a Predictable Future (lifetime)

In conventional project management, managers are always controlling the actual cost and schedule against a baseline and attempting to forecast, or predict, the estimated lifetime of the project, when the project will end, be delivered.

Schedules are tracked through what is known as Schedule Variance (SV), which can provide you with an overall assessment of all work packages in the project scheduled to date. It measures the progress of the project in dollars. Schedule variances can be determined by subtracting Budgeted Cost of the Work Scheduled (BCWS), or Planned Value (PV), which is the estimated cost of the resources scheduled in a time-phased cumulative baseline, from the Budgeted Cost of the Work Performed (BCWP), or Earned Value (EV), with is the earned value or original budgeted cost for work actually completed:

$$SV= (EV-PV)$$

Costs are tracked to what is known as cost variance (CV), which tells you if the work accomplished is costing more or less than what was planned at any point over the lifetime of the project. Cost variance can be found by subtracting the actual cost of the work performed (ACWP), or actual cost (AC), which is the sum of the costs incurred in accomplished work, from the budgeted cost of the work performed (BCWP), or earned value (EV):

Figure 4.5 shows a sample of a cost/schedule graph with variances identified for a project at the current status report date. The graph also focuses on the predictability of the expected lifetime of the project by showing two projected variables: the EAC (Estimated Cost At Completion), which includes costs to date plus a revised estimate of costs for the remaining work to be done on the project, and the BAC (Budget Cost At Completion), which shows the total budgeted cost of the baseline (project cost accounts) of the project.

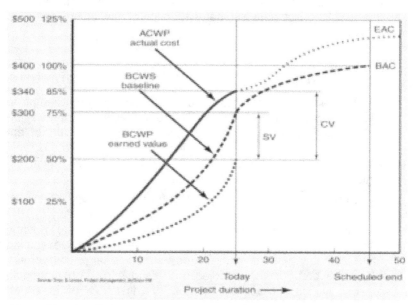

Figure 4.5 – Cost/Schedule graph provides variances identified for a project at the current status report date

Matured Technologies

Finally, conventional project management normally adopts matured technologies, already proven to work with stable results. Since the whole management approach relies on predictability and stability of project execution, the least desirable scenario will be one where a technology may not work or provide expected results.

For that reason, it is very unlikely that you will find cutting-edge technologies and approaches to conventional project management. It would be too costly to have to deal with unexpected data losses due to patchy wireless coverage, data corruption generated by virtual servers during transcontinental synchronizations, or data corruption caused by data transmission via a satellite-based car phone.

Many large organizations, particularly those in the construction industry, are still using faxes to report project data to a project assistant that will later on enter all the information on a Excel spreadsheet or MS Project. It's not that these companies don't believe in technology, or don't have the ability to implement it, but they would rather work with a system that is proven to work then take the many risks associated with early adoption of new technologies.

The APM Way: Adaptation versus Optimization

The adoption of APM should typically be an attempt to address the problem of managing large projects to improve the economics of their planning and execution. This is a critical problem that can radically benefit from an APM approach linked to the rapid emergence of the Internet and other global networks, and their role as a global disintermediated service marketplace.

Current project planning is plagued with logistical problems, such as identifying and assembling skilled teams of people, the timely procurement of products and services, and the optimal decomposition of a project into biddable pieces. Project execution almost invariably suffers from scope, schedule and cost creep. Tracking the progress of a project and recovering from schedule disruptions are *ad hoc* activities at best (the first level of the CMMI model).

The APM approach is aimed at radically changing the way teams are constructed and projects are conducted. It aims at altering the economics of providing large-scale services and at significantly improving the efficiency of implementing large projects. In APM, as discussed in earlier chapters, several of the conventional approaches to project management either do not exist or have a different role.

The technical management issues of APM provide a new paradigm for building complex projects faster and cheaper, by employing the power of distributed computing and operations. As discussed earlier, APM is not new, and does not replace conventional project management. It actually can be adopted as a way to manage critical projects with heavy resources or technical constraints, as well as instituting improvements to a conventional project management that has drifted off course.

As the demand for integrated project solutions constructed from a combination of existing and newly developed projects increases daily, many project managers find themselves with a shortage of the critical skills necessary to compete in these newly created global markets. Employing virtual collaborative project development provides a dramatic increase in a project manager's opportunities to successfully complete a project on time and under budget, but it also increases the organization's ability to compete in the marketplace by delivering fast and efficient products and services.

APM provides a broader skill and project knowledge base, coupled with a deeper pool of potential personnel to employ. It removes two of the major barriers of any project: company affiliation and physical location. APM focuses on critical characteristics, underlying how projects actually get done in traditional collocated environments, and attempts to adapt its process to a distribute operations framework. The following is a list of the main characteristics of APM:

- It can rely strongly on web-based PM
- No time for "ready," it is always on "fire..."
- Projects are "research-like" yet "mission-critical"
- Not sure of lifetime
- Risk-driven
- Breakthrough and evolving technologies
- Need for integration of speed, change and radical innovation

No Time for "Ready," It Is Always on "Fire..."

The APM environment does not allow time for ready, as the project is dynamic and tends to change on the fly. For that reason, the project execution is always on fire. The best way to characterize this is that familiar scenario when a major problem emerges during the execution of a conventional project, and a project leader or project manager paraphrases the astronauts of Apollo 13, saying "Houston, we have a problem!" From that point on it's APM. You have no idea when ready means ready, but you are on fire to produce results.

Projects are "Research-like" yet "Mission-critical"

Since APM does not seek predictability, the nature of the work is much more like a research task, yet it is still mission critical. Again, in the example of the Apollo 13 "project," there came a point, after the astronauts communicated to Mission Control in Houston that they were in trouble, and that all the conventional approaches for managing that project had failed. NASA had to bring experts into a room and tell them to innovate, to research a solution based solely on the materials and resources the astronauts had inside their capsule in space.

Nonetheless, the team of experts had a very limited and precise timeline to come up with a solution. Not only was this portion of the project on fire, but the project became very mission-critical. APM operates in this mode most of the time.

Not Sure of Lifetime

Without the basic data a conventional project management approach relies on, such as time, resources and cost, no project can predict its lifetime. With APM, the fact that it is always in a research-like mode could cause its sponsors to "pull-the-plug" on the project at any time and deprive it from its financial resources. By the same token, even if financial resources are not an issue, the lack of an acceptable solution for a particular problem could also derail the whole project.

Since virtual project managers have very little supervisory control over their staff, their role is shifted to more of a boundary manager than a supervisory one. Crossing a boundary may mean a project is dead. Having an excellent parallel execution of several tasks may mean the project will finish ahead of time. In other words, the lifetime of an APM is very difficult to predict.

Risk-driven

The challenges of successfully completing projects within the expected timeframe have increased over the last 10 years. Many project managers have relied on more effective application tools, better forecasting and the accumulation of best-practices resources, leading to continuous decreases personnel. In the "Do More with

Less" project management world, project managers have also experienced decreases in skilled staffing and budgets. Consequently, it is very common to find projects being driven by unrealistic time constraints and, therefore, delivered late. APM attempts to find a balance in planning and executing projects. Risk management is the key to finding the acceptable balance within the project management methodology.

Pareto's law, more commonly known as the 80/20 rule, is a theory about the law of distribution and how many things have a similar distribution curve. This means that typically 80% of your results may actually come from just 20% of your efforts! Therefore, realistic project planning and execution requires project teams to develop and practice proactive risk management. The vast majority of tools and applications of various project management processes stressed by PMBOK (ex. Work Break-down Structure, network logic, estimating, resource allocation earned value, lessons learned and risk analysis) are directed at reducing or removing risk impacts to a project, or capitalizing on opportunity within a project.

For all the characteristics of APM discussed earlier in this book, the agile PM has a more significant project role as "Risk Manager" than any other role. Yet, paraphrasing Professor Vijay Kanabar from Boston University, the top reasons for project failures remain virtually unchanged for the last several decades. The development of, and operating within, a proactive risk management environment, within projects and project teams, is the basic philosophy behind Risk-driven Project Management.

Pareto's law can be seen in many situations. While not literally "80/20," the principle that the majority of your results will often come from the minority of your efforts is certainly valid. So the really smart people are those who can see, upfront and without the benefit of hindsight, *which 20%* to focus on. In agile development, we should try to apply the 80/20 rule, seeking to focus on the important 20% of effort that gets the majority of the results.

If the quality of your application isn't life-threatening, if you have control over the scope, and if speed-to-market is of primary importance, why not seek to deliver the important 80% of your product in just 20% of the time? In fact, in that particular scenario, you could even question why you would ever bother doing the last 20%.

Breakthrough and Evolving Technologies

Whether or not the traditional PM processes are efficient in planning and controlling projects, the idea of incremental development is that something tangible is produced every cycle. A cycle may be from three to six weeks; it's always the same within a particular project. A regular visible delivery is good for morale: it makes everyone - developers, users, management - feel that we're definitely getting somewhere, and gives a clear idea of how the project is progressing.

Cycle deadlines are never slipped in incremental development; neither is quality reduced. Instead, goals for a cycle are dropped if necessary. This technique is called "timeboxing." In order to achieve this, each cycle must have a prioritized list of goals. The list must be a mixture of 'must haves', 'should haves', 'could haves', and 'would like to, but probably never will haves'. Because there are always lower-priority goals in every cycle, it is possible to drop goals without compromising the main track of development.

Agile Project Managers must embrace breakthrough and evolving technologies in order to mange their projects in a cost-effective and timely manner. Fueled by technological advancements, cross-functional collaborative teams and a competitive job market, alternative work practices (including virtual project teams, telecommuting and remote management of geographically dispersed employees) can help to improve the quality of project deliverables (i.e., to transmit an Excel attachment via e-mail or IM, instead of faxing the sheet), saving costs and boosting project productivity.

Actually, the growth we are seeing in the adoption of APM is being enabled by digital technology – e-mail, Web conferencing, high-speed Internet connections – but businesses wouldn't encourage APM if it didn't have bottom-line benefits. While it may not be feasible for all projects at all times, in today's global economy, proactively adopting and encouraging APM is essential.

Experts agree, however, that geographically remote project workers need both effective technology and effective communications skills to sustain a successful virtual work environment. No longer are face-to-face meetings the only way to build trust and teamwork. Armed with new technology and new best practices, APM offers new ways to connect on a human level with people anywhere, anytime. Studies like Meetings in America III show us that having the technologies is the ante to get into the APM game. Mastering how to use them is what will distinguish the agile PM winners from the losers.

To summarize this chapter, Table 4.1 provides a comparison between the main characteristics of Conventional Project Management and Agile.

Table 4.1 – Main characteristics of conventional project management versus APM

Conventional PM	APM
Project life cycles comprised of "*seven*" phases	Project life cycles loosely coupled
Process focus is more on *optimization* than *adaptation*	Process focus is more on *adaptation* than *optimization*
Client/server based	Web-based PM (distributed computing)
Built for known *functionality;* eventually there is a time for *ready*	Unsure functionality; no time for *ready*, it is always on *fire*…
Change is more of an *exception* than the norm	Projects are *research-like*, but yet *mission-critical*
There is a tendency for a predictable future (lifetime)	Not sure of lifetime
Risk-avert, conservative	Risk-driven
Matured technologies	Breakthrough and evolving technologies
Need for integration of predictable processes, standards, and best practices	Need for integration of speed, change and radical innovation

Chapter 5
A Business Case for Agile Project Management

Time to market and quality are more important than ever before.[11].

Overview

Despite the fact that project management is one of the oldest accomplishments of mankind, from the extraordinary achievements of the builders of the Egyptian pyramids, the Babylonian suspended gardens and the Great Wall of China, the great majority of companies today still do not rely on any structured form of project management. Managing by walking around is still very common; simply popping one's head in a team member's office and asking, "Hi Joshua! How's it going?" can be a very efficient and informal way to manage projects. If you keep doing this across the team a few times a month, you will find that either your project is performing well, or that you need a new team.

The problem is that in a fast-paced world such as ours, speed is everything. If we can't deliver a project fast enough, someone else in the world can, using virtual teams and virtual project management techniques.

---

[11] Siddharta Govindaraj, Silver Stripe Software Pvt. Ltd.

A Word about VPM and ePM

Although virtual project management (VPM) and electronic project management (ePM) are seen as the same, the word "virtual" makes me nervous when referring to project management. Sure, these practices try to introduce more flexibility into project management best practices, and to some extent, agility, as one can take advantage of different time zones, remote task forces and parallel crashing activities. But as a discipline, project management relies on very specific data in order to be successful, and having a virtual (if by that you mean abstract) approach to it can be very dangerous. We see no problems with virtual reality and virtual friends, but project status reports must be concrete; that is, real, not virtual. Of course, we are exaggerating the semantics of the issue here, but regardless of whether we call it VPM or ePM, what needs to happen is that information about the project must be available in cyberspace so that all teams, all members of the project, can access and update it. This should be done electronically; thus, ePM.

As figure 5.1 suggests, electronic project management is the Information Age equivalent of management by walking around. It does not replace conventional project management practices; rather, it is an evolution from it. Globalization, compression of product life cycle, global competition and many other factors have contributed to a shift in project management to a more ubiquitous and pervasive approach. Recent Internet-based collaboration tools have also contributed to new possibilities for web-based project management. Thus, understanding 21st century project management challenges, and how ePM can support and mitigate these challenges, can be very advantageous to any project manager.

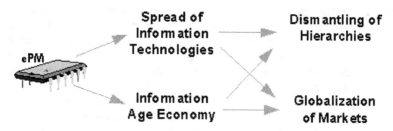

Figure 5.1 – APM is the information age equivalent of management by walking around

The Globalization of Markets and Projects

The fast-paced, growing integration of economies and societies around the world has been one of the most hotly debated topics in international economics over the past few years. In order to be competitive in the 21st century global and knowledge economy, organizations must be able to provide cost-effective products and services, while maintaining their level of quality (if not exceeding it). Hence, management of virtual projects becomes a very important factor for the success of such ventures, as it enables project managers to better assess and control projects from a distance.

Globalization has enabled rapid growth and poverty reduction in countries such as China and India, among many others that were poor 20 years ago. But globalization has also generated significant international opposition over concerns that it has increased inequality among some nations, and environmental degradation to others. The main opposition, however, has been toward the sharp increase of project outsourcing, particularly by the United States. Organizations are moving much of their production activities overseas, taking advantage of lower-cost labor markets, favorable taxation laws and highly skilled professionals.

We believe this trend, in the long run, will be beneficial to the world economy as a whole, as much as when blue collar jobs in the United States were outsourced overseas, thus generating white collar jobs. But to be successful these projects still need to be well-managed, regardless of geographic distances, cultural diversity and time zone limitations. By having ePM and APM expertise, project managers will be equipped to deal with the unique challenges of managing virtual teams and promoting agile project management execution, while handling cultural and political issues that are often foreign to their practice. And it is important that when we talk about the challenges that ePM and APM represent to project managers and executives, not only technically, but in terms of financing, we should also recognize, and be brave enough to admit, that project development and execution is not just a question of economics; it is also a question of belief, morality and ethics; it is a question of spirituality. Because wherever your project is, or wherever you decide to build your virtual team, be it in Brazil, Argentina or India, be it in slums or in villages throughout the world, you find the importance of spirit, the importance of culture, the importance of the environment in which people live. We are not sure what the next collar color will be after the outsourcing of the white (how about grey?), but we are certain that it will only take place if projects are well-managed and successful.

When, according to the Project Management Institute (PMI), the average project tends to experience 168 percent cost overrun, (and of these, only 22 percent are successful - projects that finished on time and under budget), who is to blame? Certainly not the sponsors and project managers, for their decisions to push for cost-cutting measures (one of the main reasons given for outsourcing). Certainly not the fact that we do not have enough qualified professionals with the right expertise, particularly for IS/IT-oriented projects (another reason given for outsourcing practices).

The reasons for such discouraging statistics are many, and while globalization and outsourcing are at the core of most of these reasons, a fact which goes beyond the scope of this book, the lack of good project record keeping and effective measuring of project performance and results are, in my view, at the core of the problem. According to Microsoft, more than 60 percent of project managers are still using Lotus/Excel spreadsheets to manage and keep track of their projects. Although these applications do a great job in tracking results, they are a far cry from the minimum set of features a project manager needs to adequately manage their projects, as we will discuss later in this book.

These challenges have led to the emergence of quality movements across the world, with ISO 9000 certification a requirement for doing business. ISO 9000 is a family of international standards for quality management and assurance that covers design, procurement, quality assurance and delivery processes for everything from banking to manufacturing. Of course, quality management and improvement invariably involve project management.

The Rise of Third World and Closed Economies

There has been an ever-increasing demand for managers of consumer goods and infrastructure development since the fall of the Soviet Empire and the gradual opening of the Asian communist markets. As western companies rush to establish a presence in, and capitalize on, those markets, many are taking advantage of project management techniques to develop supply chains and establish distribution channels and foreign operations in those countries.

Such rapid development of third world markets and closed economies will continue to increase in geometric perspectives, pushing the adoption of ePM to much higher levels than we see today. Saying that India had "one of the most closed economies of the world," US Trade Representative Robert Zoellick has called on more competitive developing countries such as India, China and Brazil to open their markets in order to sustain support in the United States and elsewhere (http://www.caltradereport.com, 3/15/2004). Zoellick defends that more competitive developing nations such as Brazil and Argentina should not expect to get the same kind of special and preferential treatment — such as longer implementation schedules — as the poorest countries get in agriculture and other sectors. Nonetheless, these economies are very attractive to outsourcing due to their highly skilled professionals, affordable labor, and manageable time zone restrictions.

For instance, at MGCG, 80 percent of its systems development projects are executed in Argentina. Although the strategy is a form of outsourcing, we do not consider it as such, as we have a presence in Buenos Aires, as well as Brazil. From the start, our tactic was to incorporate this skill set from abroad into our internal corporate operations, allowing them to be an integral part of the process and project management. By tapping into ePM and APM techniques we can pull those resources in, transparently, to the whole organization. To the staff, and mainly our clients, our information systems and technology (IS&T) group appears to be in the next building, or across town, and not thousands of miles away, except early in the morning or late afternoon, during winter months, when the time zone difference is three hours. But that is not a problem for us, as in the US we have been accustomed to the consistent three hour difference when dealing between the West and the East coast. For the most part, working with our Brazilian and Argentinean counterparts is actually more productive than working cross-country in the US, as for the most part the time difference is of only one to two hours, for the greater part of the year.

The reason many outsourcing projects fail is precisely due to the fact that these outsourced tasks are decoupled from the overall process management of the project, and do not take into consideration APM best practices, as discussed in the earlier portion of this book. We believe much of this happens due to the poor levels of measuring and controlling tools and practices available, or known, in managing those virtual projects, as well as a poor understanding of SCRUM (no wonder certification is in demand!).

Corporate Downsizing

As we traverse the meltdown of the world's economy, General Motors (GM), which in 2000 had laid off about 20 percent of its Wilmington, Delaware assembly plant due to poor sales of some of their new automobile models, has filed for bankruptcy during the development of this book, in the spring of 2009. While they reorganize under chapter 11, agile project execution would be of the essence for their success. The longer they stay under chapter 11 protection the worse it will be for their business.

A company cannot carry on constant growth without proper management. What this means is that at some point, companies must take time out of their hectic schedules and start crossing out names of employees they can do without. Despite the fact that you may be an excellent employee, there is never a guarantee that you can make it past the downsizing/rightsizing of many companies. You may not lose your job, but you may become responsible for double your original tasks, and executing those tasks with *agility* and effectiveness becomes very important! Nonetheless, the reorganization of a company's personnel layout does not simply involve firing employees; it also denotes opportunities that can benefit workers with increased responsibility, salary and prospects for future advancement. Again, APM can minimize this Darwinist approach by enabling professionals to be engaged on several projects fast and effectively, with reduced personnel and time.

The last fifteen years has seen a remarkable restructuring of organizational life and business processes. These business process reengineering (BPR) and business process improvement (BPI) projects generated a chain of corporate downsizing (or rightsizing, if you are still employed), calling for a return to the core competencies necessary for the survival of many organizations. It was then that the middle manager disappeared and began to be replaced by project management as a strategy to ensure that tasks were getting done. Organizations became flatter and leaner, while realizing that change was inevitable and, therefore, had to be managed.

It was also then that, as a result of downsizing, companies began to outsource, promoting a major paradigm shift to the way projects were managed. More than ever before agile project management techniques and best practices became vital to a successful project execution. The challenge - and we are still dealing with this today, more than ever before - was that companies were outsourcing a significant segment of project work, meaning that the project manager had to manage not only their own staff, but also their counterparts in different organizations, often outside the company, the building, the state or the country.

Effects of Corporate Downsizing in Agile PM

The long-term effect of corporate downsizing has been a change in the way projects are managed, especially when downsizing equates to mergers. In my last major project, with Virtual Access Networks, we successfully completed phase one when our product was launched at Comdex, Fall 2001, and awarded the Best Enterprise Product. Therefore we thought our project was going to enter phase two, as we became profitable and were getting ready to expand the business. Instead, the company was acquired and as a result of the merger, the new established strategy morphed the project into the buying company current product. In other words, the project was killed, and the members of the team abruptly dispersed in the name of cost-efficient operations.

The competitive global marketplace is actually making sure that downsizing is here to stay. As a result, companies have to run their businesses with efficiency and cost-effectiveness, and project management - virtual project management, as most projects are no longer local to organizations - is an important technique in this process. Companies are no longer downsizing due to general economic conditions, but for better staff utilization, outsourcing, plant closures, mergers, automation, and the use of available new collaboration and communication technologies.

The adoption of APM is a strong business justification to mitigate corporate downsizing. APM teams can be formed quickly and are agile in nature. They can help organizations decrease their response time to changes in today's hyper-competitive markets. Organizations are also able to leverage expertise that is dispersed over geographic areas and was previously left untapped. Furthermore, virtual teams can also benefit employees as they lessen the disruption of the employee's life by requiring less travel time to meet with dispersed teams. Additionally, team members have the opportunity to broaden their experience by working across organizations and cultures.

As with the Virtual Access Networks example, professional individuals are also affected. It is a sad scenario when projects need to be shut down due to poor performance, in particular when excellent professionals are let go. Research shows that being let go is likely to mean the loss not only of upward career movement, but also of economic stability and self-respect. As a management consultant, I have seen this situation all too many times: projects are killed, and the members of the team feel like they are being sacrificed, not because their projects were in serious financial trouble, but rather because the cost performance index (CPI), or profits being made, were not high enough.

Another consequence that many project management professionals take with them after a failed project is the low morale among team members left after the disengagement or downsizing process has been addressed. Companies that help their project team get new jobs, and provide outplacement services, end up much better positioned than companies which simply wield the ax. This is because they have a better chance of retaining the loyalty of the surviving professionals. This is particularly true in virtual project management, as we'll discuss in further chapters. Trust is one of the most valuable yet brittle assets in any project, especially with ePM, when you don't have the level of interaction you do with conventional project management.

Project managers should remember that they are artificial creatures chartered by the sponsors and the stakeholders. As such, they are subject to their values, and their approvals can be withdrawn at any time. Hence, efficiency and cost-effectiveness is becoming paramount as never before. The need for full implementation of agile project management is becoming a must.

APM Can Mitigate Project Downsizing

In our practice, we have been advocating the use of APM techniques as an alternative to project downsizing and change of scope to control costs. In particular, APM can address problematic projects on several levels, including the:

1. Determination by project sponsors to downsize or dramatically reduce the scope of the project

2. Mitigation of morale problems affecting the project after the scope reduction or downsizing decision

3. Development of APM procedures for reallocation of resources with a minimum impact to the current project Organization Breakdown Structure (OBS) and The Work Breakdown Structure (WBS)

4. Long-term effects of the project's downsizing or change of scope efforts.

APM Mitigates Scope Changes

Today's project managers are under ever-increasing pressure to deliver results – in the form of projects that drive improvements to the bottom line – even while project scope and budgets are being significantly slashed. Meanwhile, despite the fall of the Internet economy, business environments continue to change at a rapid pace, leaving many project organizations struggling to keep up with the pace of change. These changes have led to an increased interest in APM methodologies, with their promise of rapid delivery and flexibility while maintaining quality.

APM methodologies such as eXtreme Programming (XP), SCRUM and Feature-Driven Development strive to reduce the cost of change throughout project execution, especially the software development process. For example, XP uses rapid iterative planning and development cycles in order to force trade-offs and deliver the highest value features as early as possible. In addition, the constant, systemic testing that is part of XP ensures high quality via early defect detection and resolution.

Typically, project sponsors tend to downsize or reduce the scope of projects to cut cost, improve efficiency and maintain a CPI level acceptable to their stakeholders and shareholders. Such phenomenon in the United States and foreign countries is, unfortunately, here to stay. Every day we hear and read in the news of projects that are being abandoned or having their scope dramatically reduced for many reasons, but always pointing to the financial aspects of the project. But there is a difference between reducing scope and terminating projects, and it is important to understand those differences if one attempts to mitigate those risks and even rescue a project.

The reduction of a project scope, if not for a technical reason, usually refers to a deliberate decision to reduce the amount of resources, or work force, that is intended to improve project performance. We, at MGCG, view such an approach as part of a larger plan, where sponsors analyze their core business and develop their business to its fullest extent. In other words, the organization often goes back to the basics, making the scope of project suddenly no longer desirable. When reduction of project scope does need to occur, we recommend sponsors and project managers to look at whether such reduction of scope is proactive, which is a restructuring strategy of a project to obtain efficiency and market share, or reactive, where the action occurs because of deep financial trouble within the project. In my view, reduction of project scope, if not for technical reasons, occurs because of:

1. An intentional decision by project managers and/or sponsors

2. A reduction of project resources, personnel, often disproportionately in the management ranks

3. Efficiency and/or effectiveness objectives

4. Changes in work processes, often caused by dangerous delays in schedule or budget overruns.

In spite of some early success with agile methodologies, a number of factors are preventing their widespread adoption. When we carefully consider the itemized list above it becomes clear why agile methodology advocates often find it difficult to obtain management support for implementing what seem like dramatic changes in project management. These methodologies require project workers, managers and users alike to change the way they work and think. For example, the XP practices of pair programming, test-first design, continuous integration and an on-site customer can seem like daunting changes to implement. Furthermore, these methodologies tend to be developer-centric, and seem to dismiss the role of project managers in ensuring success.

Again, it is no wonder that the whole theme of reduction of project scope usually equates to lowering operating expenses, mostly direct costs, where crashing proved not to be effective. After all, in recent years emphasis on time-to-market has taken on new importance because of intense global competition and rapid technological advances. The market imposes project duration date. Such analysis becomes even more serious during recession periods, such as the one we are experiencing in the United States and throughout the world, when cash flows are tight.

Other reasons to reduce project time or scope occur when unforeseen delays, caused by adverse weather, design flaws, equipment breakdown, etc., cause substantial delays midway in the project. The adoption of APM in many of these cases can be an excellent alternative to reduce cost, improve efficiency and realign project goals. In our practice, we have found that strong management is absolutely critical to the successful adoption and application of agile methodologies. But we have also discovered a lack of alignment between the methodologies and tools of traditional project management, and those of newer agile methodologies. In addition, we believe this misalignment points to a deeper problem; differences in fundamental assumptions about change, control, order, organizations, people and overall problem-solving approach. Traditional management theory assumes that:

- Employees are interchangeable "parts" in the organizational "machine"
- Hierarchical organizational structures are means of establishing order
- Increased control results in increased order
- Organizations must be rigid, static hierarchies
- Problems are solved primarily through reductionist task breakdown and allocation
- Projects and risks are adequately predictable to be managed through complex upfront planning
- Rigid procedures are needed to regulate change

Within this context, it is small wonder that the new methodologies appear informal to the point of being chaotic, egalitarian to the point of actively fostering insubordination, and directionless in their approach to problem solving. We believe that the slow adoption of agile methodologies stems mainly from this misalignment between the fundamental assumptions of traditional management and those of the new agile development methodologies. As such, we believe there is a significant need for a change in assumptions and a new management framework when working with agile methodologies. Countless projects have seen the negative effects of downsizing or reduction of scope. Morale suffers, which equates to lower productivity and cost-efficiency for the project. Project managers tend to complain of the morale-sapping character of most downsizing efforts and reductions of scope, and how low morale creates anxiety and paralysis within their project teams to the subsequent detriment of productivity. When a professional loses his/her job, confusion and anxiety sets in. Many of them are thrown into unexpected transitions with extended middle periods, characterized by confusion and a lack of ability to move forward. Such phenomenon can have long-term consequences for the project at hand.

Most importantly, in our 20 years of experience managing projects, both in the United States and abroad, we have found that a 10 percent reduction in project personnel actually resulted in only a 1.5 percent reduction in costs. In addition, we have found that a project worker's trust and empowerment were shattered after the downsizing or significant reduction of scope for non-technical reasons. This caused workers that stayed after the restructuring of the project to show less initiative in getting the work done. Their feeling was that they would be the ones terminated next, so their attitude was "Why try to do the job since I am going to be the next one laid off?"

Again, APM practices can help in this process by allowing a more creative way to deploy human resources, such as using SCRUM methodologies, where a significant amount of money can be saved. Another way is through the use of XP teams, with very focused and specialized know-how, without the overhead cost of engaging them full-time or relocating them to the project's site.

Agile PM Framework for Reallocation of Resources

In the search for a new framework, we have come to believe strongly in emerging management principles, based on an approach of complexity, that exploit an understanding of autonomous human behavior gained from the study of living systems in nature. Specifically, we believe APM can be an adaptive tool to our management assumptions and practices. Like APM teams, project managers also need a set of simple guiding practices that provide a framework within which to manage, rather than a set of rigid instructions. Following these practices, the manager becomes an adaptive leader – setting the direction, establishing the simple, generative rules of the system, and encouraging constant feedback, adaptation, and collaboration. This management framework can provide project teams implementing agile methodologies with:

- A recognition of the limits of external control in establishing order, and of the role of intelligent control that employs self-organization as a means of establishing order
- A view of organizations as fluid, adaptive systems composed of intelligent living beings
- An intrinsic ability to deal with change
- An overall problem-solving approach that is humanistic in that:
  - It limits upfront planning to a minimum based on an assumption of unpredictability, and instead lays stress on adaptability to changing conditions.
  - It regards employees as skilled and valuable stakeholders in the management of a team.
  - It relies on the collective ability of autonomous teams as the basic problem-solving mechanism.

Traditional project management methodologies grew out of a need to control ever-larger development projects, and the difficulties of estimating and managing these efforts to reliably deliver results. These methodologies drew heavily on principles from engineering, such as construction management and the Software Engineering Institute (SEI) Capability Maturity Model Integration (CMMI). As a result, they stressed predictability (one must plan every last detail of the bridge before it can be built), and linear development cycles – requirements led to analysis which led to design which in turn led to development. Along with predictability, they inherited a deterministic, reductionist approach that relied on task breakdown, and was predicated on stability; stable requirements, analysis and design. Techniques such as Performance Evaluation and Review Techniques (PERT) and Critical Path Methods (CPM) became a must-have approach to project planning and execution.

This rigidity was also accentuated by a tendency towards slavish process "compliance" as a means of project control. When a project manager decides to downsize or reduce scope, a detailed and well-planned program has to be implemented to make it successful. Unfortunately, far too many project managers restructure their projects without looking at the negative repercussions of their effort, regardless of whether the order came from sponsors or not. Some project managers are able to increase profitability and/or productivity of their projects, but for many restructured projects the results have been disappointing. We have found a few reasons why project downsizing or reduction of scope does not work in many projects:

1. Project managers fail to set up retraining polices and do not foresee the human resource problems that develop when downsizing/scope reduction takes effect. APM can effectively reach out to project teams and mitigate these adverse results through SCRUM methodologies.

2. As the project is downsized/scope changed, the added burden of work is thrown to line or project leaders with limited time to do the job and limited expertise in that field. APM is ideal for such situations.

3. As projects become flatter, the past communications that worked well are gone. No communication link between the different levels within the organization takes place.

4. As mentioned earlier, quite often productivity and quality suffer because there is no plan for how the work is to be done with a reduced staff.

5. A project restructuring eliminates the job security, the sense of belonging. When this happens, project workers show no loyalty to the project and will move from firm to firm for greater pay. The mentality then, is like a "gun for hire".

While these methodologies may have worked for some organizations in the past, and may still work in some circumstances, for many companies these methodologies only added cost and complexity, while providing a false sense of security that management was "doing something" by exhaustively planning, measuring, and controlling. Huge costs were sunk in premature planning, without the rapid iterative development and continuous feedback from customers that we have come to realize are prerequisites for success today.

The results are bleak, packed with repeated, public failures such as the London Ambulance System and the Denver Airport Baggage system, to name a few, which earned the software industry a reputation for being "troublesome," with huge cost overruns and schedule slippages. Consider the results of the Standish Group's CHAOS surveys. In the first survey, it was estimated that only 18 percent of all software projects were considered successful, 31 percent were failures and 53 percent were challenged. Comparatively, the 1998 figures showed a marked improvement, in which 26 percent were successful, 46 percent were challenged and 28 percent were failures. The study attributed the increase in success to scaling the size of projects back to manageable levels using smaller teams. This result is clearly in line with the principles of agile methodologies.

Therefore, for project managers to have a successful program in any downsizing/scope reduction efforts, various points will have to be addressed. The development of a well-detailed plan to train each retained worker to do the work is a major one. Doing this will prevent rehire "creep" from taking place. In other words, hiring consultants or rehiring project workers to do the work once done by the laid off staff will not be mandatory. Another part of making the restructuring program a success is keeping the severance-related cost to a minimum. To do this, the project manager should use non-voluntary programs that identify who will be affected, rather than using a voluntary program. When voluntary programs are implemented, they are often overused by people who have high severance packages.

Failure to attend to this transition with caution and open communication can have tremendous impact on a project. As a project leader for a bridge construction in Brazil back in the early 80's, I witnessed the project manager going through a very hard and uncertain time for failing to observe the points outlined above (unfortunately APM was not an option back then). Shortly after the project reorganization, we started to lose employees because someone decided we didn't need as many people. No one seems to know how they figured out how many employees we needed, and we simply did not have enough people to do the job that was demanded of us. This was not a good place to work anymore. When the right job comes along, and it will, I'll leave my current position.

In summary, it is apparent then that "getting smaller" is not enough when it comes to project management. Downsizing or reducing scope is the equivalent of project anorexia, as it can make a project thinner yet not necessarily healthier. Unfortunately, there are times that such restructuring becomes necessary, especially in cases of technical constraints or market shifts. Nonetheless, project managers should always proceed with caution when contemplating restructuring of their projects, and look at the relationship between restructuring the project and productivity, as well as project management effectiveness after the fact. Just because the project is now leaner and the critical path has been crashed, does not necessarily mean it is more efficient and cost-effective. Research also shows that over the long term, the financial health of many projects diminishes with the downsizing/crashing efforts, especially for long-term projects and those not suffering from technical constraints. Some projects increase profitability and productivity, but for many, downsizing/reduction of scope has been disappointing. Furthermore, we have found that many established project management practices still apply to agile development projects – with some adaptation and a strong dose of leadership.

Chapter 6
Trends, Camps and Collisions

Traditional processes are too rigid to address today's project management concerns.

Overview

While managers designed traditional methodologies in an effort to control projects, the technical community gave birth to agile methodologies in response to their frustrations with traditional management (or lack thereof), and the resulting collision of their methods, products, and morals have generated a few trends and camps. For example, the principles of XP are focused almost entirely on the development process. While the technical community has championed these principles, very little has been written about the management side of agile development projects. The implication is that there is little need for a project manager since XP teams develop and monitor their own tasks. No wonder that corporate management has been skeptical of agile methodologies and slow to embrace them.

At the same time, Project Management Institute (PMI) membership has been growing fast, and the demand for Project Management Professional (PMP) Certification is on a near vertical ascending demand. PMI's Project Management Body of Knowledge (PMBOK) guideline is now in its fourth version and has become a de facto standard in the industry, establishing a strong camp in the area of project management best practices. Meanwhile, in another camp, managers conjure up an image of a room full of developers doing their own thing.... and the name "eXtreme" doesn't help matters either.

Regardless of the particular trend, camp, or methodology, the traditional project manager is often seen as a "taskmaster" who develops and controls the project plan that documents (in excruciating detail, under the PMBOK) the tasks, dependencies, and resources required to deliver the end product. The project manager then monitors the status of tasks and adjusts the plan as necessary. Underpinning this mechanistic approach is the assumption that equates individuals to interchangeable, controllable commodities.

So for many managers comfortable with traditional methodologies, the prospect of implementing agile methodologies in their development projects can be daunting. But it doesn't need to be. In fact, independent of agile methodologies, other trends in project management point to a convergence between the management community and the technical community. But before we can talk about this convergence, let's review the major camps and trends, beginning with the PMBOK.

The Project Management Bodies of Knowledge

The Project Management Bodies of Knowledge (PMBOK) have now been around for more than fifteen years. They have come to represent one of the most important features of project management. They are used by many organizations as the basic template upon which project management competency is defined and assessed.

But not many project management professionals, except maybe those that are PMP certified, know what a Body of Knowledge (BoK) is, why it is important, how a BoK differs from a Body of Competence (BoC), how many BoKs there are, what the issues are in creating a valid (universal) BoK, and what the British Association for Project Management (APM) is doing in this area. In addition, we believe it is important to understand the role of BoKs in project management, so that we can better rely on them for APM. PMI is extensively researching the subject and routinely adding the topic of managing virtual projects to its best practice.

What is a BoK?

All the project management societies' BoKs, except PMI's (as seen in figure 6.1), are misnamed: they are not bodies of knowledge, as such, but guides to the topics that a project management professional ought to be knowledgeable in, to some degree at least. APM's BoK simply describes the nearly 40 elements that it believes a project management professional should be knowledgeable in. To be precise, it describes three different levels of knowledge for each of these elements. The APM BoK was developed for candidates for the Certificate Project Manager (CPM) qualification to assess their level of knowledge in preparing for the CPM qualification. It is now used by APM as a general normative guide; for example, in accrediting courses, in its Continuing Professional Development (CPD) program, and now as the basis of its Association for Project Management Professional (APMP) qualification, the equivalent of PMI's Project Management Professional (PMP) certification.

Figure 6.1 – PMI's PMBOK fourth edition, issued in the Fall of 2008

The BoK then is not an all-encompassing textbook in which all project management knowledge is defined. It is doubtful, in fact, if such a thing could be written (though we might get close to it). For this reason PMI has termed its version, the "Guide to the Project Management Body of Knowledge".

Why is a BoK Important for APM?

Almost every project manager is interested in discovering what they need to know about project management to be considered a competent professional. They look to the professional societies to tell them. And typically they value a certificate or qualification that says they have attained a required standard of knowledge. Many organizations use the BoK as the project management component of their more general, company-specific definition of competencies. The same is true for APM via the Scrum and ScrumMaster Certification (CSM). To know that they have been certified under the same standards you abide by becomes even more important. Figure 6.2 provides a screenshot of the ScrumAlliance website, where you can find all the necessary information to become a CSM.

Figure 6.2 – The ScrumAlliance website, source of CSM certification information and knowledge

In our project management practice in Brazil, eighty percent of our project staff is PMP certified. Despite language and cultural differences, it is comforting to know they are all certified by PMI under the same BoK (PMBOK); as in APM, we do speak the same language.

BoKs and Body of Competencies

Competencies are the things one needs in order to perform a function competently. The key point about competencies is that they are based on measures of the ability to do a task. They relate primarily to how one performs. They say, for example, that in order to perform effectively as a project management professional one might need to have knowledge, skill and experience in certain technical and commercial areas, have certain personal qualities and organizational skills, and have such and such knowledge of the organization – and be experienced, knowledgeable and skilled in project management.

The difficulty comes in defining what the areas of project management knowledge should be. Almost always the actual definition of competencies is not just company specific but role specific. Companies find the professional BoKs (APM's, PMI's, etc.) useful because they provide an authoritative international template that will help them answer this question.

The guides to knowledge themselves do not represent competencies. Competence is only demonstrable by being able to prove an ability to perform in a given role. The nearest the BoKs come to this is when a candidate demonstrates that their level of knowledge, experience, and skill qualifies them to be recognized as having attained either a company-specific standard or a national/ international standard, such as a PMP, APMP, German Fachman, or CPM.

How Many BoKs are there?

To know how many BoKs there are may not be relevant to the local execution of projects, where all professionals are co-located. However, being aware that there are more than one BoK, that PMI's BoK is not the only one, will also make you aware of the fact that some international project management professionals may have different approaches or techniques in executing a project, aside from technical terms that may be known by different definitions. For instance, PMI's PMBOK defines the amount of time a task can be delayed without delaying the project completion date as a float. Other BoKs may define float as slack. So, when interacting with project managers that do not have PMBOK's knowledge, you may find that some of them may not know what you mean by float, even though they do understand the concept of slack, and vice-versa. The same goes for project initiation versus project definition, etc.

The oldest and, in number of people and countries affected by it, the most influential, of the BoKs is PMI's. Next in terms of people and countries comes APM's. There are no other significantly different BoKs, though the French Project Management Institute uses a scaled-down version of the APM BoK, and the German and Swiss use their own modified versions of the APM's. The Dutch basically use APM's BoK. The Australian Institute for Project Management uses PMI's. Many national chapters of PMI of course use PMI's. IPMA, the International Project Management Association, is currently working on an attempt to harmonize the BoKs of its member associations.

Issues in Creating a Valid (Global) BoK

In our view, the creation of a global BoK is vital for APM, as it would standardize the taxonomy and practices of project management across the globe. There are, however, a number of challenges in creating such a global standard, particularly:

- What should the elements be?
- What is a proper definition of each of these elements?
- How should the elements be structured?

PMI has 47 elements divided into three levels of a Product Breakdown Structure. APM has 44 at two levels. The APM set is very much broader than PMI's. Fifteen of APM's elements represent all 44 of PMI's. There are three core issues in defining the elements. Agreeing whether the PM professionals remit begins before the project definition is agreed, and indeed how far into operations it should go. Deciding how much the PMBOK should reflect the broad range of issues, skills and knowledge – the management of technology and design for example – that a project management professional will encounter in successfully managing a project.

Deciding to what extent these BoKs are industry specific is another issue of global perspective. Deciding what actually the elements are in any professional (guide to a) PM Body of Knowledge is critically important. We are still a fair way from having accomplished this. Deciding authoritative definitions is equally hard. In fact at times it is very hard.

Language is another issue, as it varies not just between nations (even those that speak English), but between industries as well. Some industries have different conceptions, for example, of what words like systems engineering, configuration management, procurement, mobilization, and logistics mean. One of the challenges for the BoK writer, therefore, is to strike a balance between creating a genuinely useful general language of project management and putting people off with unfamiliar words.

For some people the first discussion regarding a BoK is its structure. Indeed there have been many debates as to the relative merits of PMI's process-based BoK over APM's four functional columns. Actually, while structure is very important, if only from a communicative viewpoint, logically it is not central. Once the elements and their definitions are agreed upon, they can largely be structured and restructured as required. This is not to say that discussion of the structure is not important. It is. But it can be disconnected from the discussion about content.

The Future

The outlook for really good work in this area is now very encouraging. PMI has been working on updating its Guide to the PMBOK. In doing so it is paying explicit attention to other work on the BoK, such as that done at the University of Manchester Institute of Science and Technology (UMIST), the International Project Management Association (IPMA), in Australia and in Holland.

There is increasing recognition that this kind of normative research needs to be extended to give better guidance on what best practice is, or rather, what best appropriate practice is, especially for the global perspective. Many companies want to pick up the best of PMBOK practice and, where it is appropriate, emulate it. Work on such Best Appropriate Practice is now beginning seriously.

Before long we might, with luck and a lot of hard work, have some empirically validated guides to what truly constitutes good, and best, practice in the management of projects, what the best writing and teaching has to say, and where help can be found. All of this, internet-based, at the touch of a button; at the heart of it will be the BoK. We are, we believe, about to enter the second generation of project management Bodies of Knowledge: a maturing of project management. PMI is already featuring a lot of APM resources at their site, as illustrated in Figure 6.3.

Figure 6.3 – PMI has several agile project management resources available at their website.

Trend: How Much Your Time is Worth

In order for APM to be successful, or any project management, for that matter, you should always focus on results. Now, the first part of your focus on results should be to work out how much your time, and that of your peers and your team, costs. This helps you to see if you, and your project team, are spending time profitably. As you work for an organization, calculate how much you cost it each year for that project. Include your salary, payroll taxes, the cost of office space you occupy, equipment and facilities you use, expenses, administrative support, etc. If you are a self-employed project manager, work the annual running costs of your business.

To this figure add a 'guesstimate' of the amount of profit you should generate by your activity. If you work normal hours, you will have approximately 200 productive days each year. If you work 7½ hours each day, this equates to 1,500 hours in a year. From these figures, calculate an hourly rate. This should give a reasonable estimate of how much your time is worth; this may be a surprisingly large amount!

Next time you are contemplating whether or not to take a task on, think about this value - are you wasting your, or your organization's, resources on a low-yield task? Are there tasks that can be automated? Calculating how much your time is worth helps you to work out whether it is worth doing particular jobs. If you have to spend much of your time doing low-yield jobs, then you can make a good case for employing an assistant, or investing in technology to automate some of those tasks.

Improving Time Management with APM

Time management skills are essential for successful project managers and workers. These are the practical techniques which have become a trend in helping the leading people in business, sports and public service reach the pinnacles of their careers. The same is true in APM. Professionals who use time management techniques (and technology), and take advantage of the world's different time zones, are routinely the highest achievers in all sorts of projects execution. Time management is essential for a successful APM strategy. If you use these skills well, then you will be able to function effectively, even under intense pressure. At the heart of time management is an important shift in focus brought to you by APM: concentrate on results, not on being busy.

Many project managers and workers spend their days in a frenzy of activity, but achieve very little because they are not concentrating on the right things.

The 80:20 Rule

Vilfredo Pareto, an Italian economist, "discovered" this principle in 1897 when he observed that 80 percent of the land in England (and every country he subsequently studied) was owned by 20 percent of the population. Pareto's theory of predictable imbalance has since been applied to almost every aspect of modern life. Simply put, as depicted in Figure 6.4, the 80/20 rule states that the relationship between input and output is rarely, if ever, balanced.

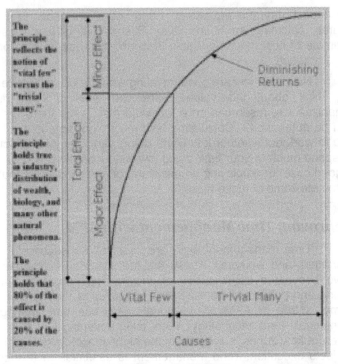

The principle reflects the notion of "vital few" versus the "trivial many."

The principle holds true in industry, distribution of wealth, biology, and many other natural phenomena.

The principle holds that 80% of the effect is caused by 20% of the causes.

Minor Effect

Major Effect

Total Effect

Diminishing Returns

Vital Few | Trivial Many

Causes

Figure 6.4 – The Pareto Principle, or the 80/20 rule, was named by the statistician J. M. Juran, in the 1940's, after Vilfredo Pareto (1848-1923).

When applied to projects, this means that approximately 20 percent of your efforts produce 80 percent of the results. Learning to recognize and then focus on that 20 percent is the key to making the most effective use of your time. As depicted in figure 6.4, the '80:20 Rule' argues that typically 80% of unfocussed effort generates only 20% of results. The remaining 80% of results are achieved with only 20% of the effort. While the ratio is not always 80:20, this broad pattern of a small proportion of activity generating non-scalar returns recurs so frequently as to be the norm in many areas.

Here are some signs that will help you to recognize whether you're spending your time as you should:

• You are in your 80 percent if the following statements ring true:

  • You're working on tasks other people want you to, but you have no investment in them.

  • You're frequently working on tasks labeled "urgent."

96

- You're spending time on tasks you are not usually good at doing.
- Activities are taking a lot longer than you expected.
- You find yourself complaining all the time.

- You are in your 20 percent if:
  - You're engaged in activities that advance your overall purpose in life (assuming you know what that is — and you should!).
  - You're doing things you have always wanted to do or that make you feel good about yourself.
  - You're working on tasks you don't like, but you're doing them knowing they relate to the bigger picture.
  - You're hiring people to do the tasks you are not good at or don't like doing.
  - You're smiling.

To Clone or Not to Clone?

In the movie Multiplicity, Doug Kinney (Michael Keaton) is an overworked professional that allows himself to be cloned so that he can accomplish collectively what he can't do individually. It doesn't work out. But with APM you might be able to pull it off. Many tasks can be automated and several people can be working in parallel, as long as, much like a conductor, you can keep the orchestra in sync.

Nonetheless, if not planned well APM can backfire and won't work, much like in the movie. Technology analysts predict something similar for the typical office worker, only this time multiplicity pertains not to people but to the gizmos we all rely on. By 2007, says Meta Group Inc. analyst Steve Kleynhans, a knowledge worker will depend on at least four different devices: a home PC, a corporate computer, a mobile information device, and a "smart digital entertainment system."

As in our example from MGCG, illustrated in Figure 6.5, the idea of equipping APM workers with multiple gadgets for ePM or APM may sound expensive, but Kleynhans claims that if companies study the options coming into the market, they will be able to reduce overall information technology (IT) expenses and make workers more productive. Today Microsoft, through its .Net Platform, enables a multitude of devices and platforms to be integrated over the web, facilitating the process of information exchange.

Figure 6.5 – Microsoft's .Net Platform.

Agile PMs as Visionary Leaders

The best Agile PMs aren't just organizers; they combine business vision, communication skills, soft management skills and technical savvy with the ability to plan, coordinate, and execute. In essence, they are not just managers – they are leaders. While this has always been the case, agile project management places a higher premium on leadership skills than ever before.

For example, XP teams create and monitor their own iteration plans in collaboration with customers, and attempt to envision a strategy to increase effectiveness, such as turning a sequential process into an overlapping one (as illustrated in Figure 6.6). The customer creates stories (features) and prioritizes them based on business value. The developers divide up the tasks themselves, as they work, and measure progress for each iteration (time-boxed development cycle), adjusting plans with the customer as necessary. So, if the project no longer needs a detailed master project plan, why does it need a project manager? Because every project needs a leader.

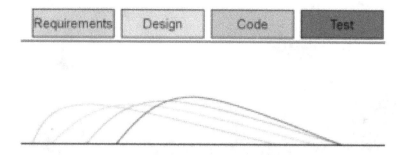

Figure 6.6 – Agile PMs act as visionaries as they envision the transformation of sequential processes into overlapping ones.

Agile methodologies free the project manager from the drudgery of being a taskmaster, thereby enabling him to focus on being a leader, someone who keeps the spotlight on the vision, who inspires the team, who promotes teamwork and collaboration, who champions the project and removes obstacles to progress. Rather than being an operational controller, the project manager can become an adaptive leader – if she can relinquish her reliance on old style management.

The basic phases of an agile development project are really no different from those of any other project. You still must define and initiate the project, plan for the project, execute the plan, and monitor and control the results. But the manner in which these steps are accomplished is different, and requires the project manager to retrofit what they know about traditional management to a new way of thinking – the thinking of complex adaptive systems.

For instance, according to Harvard Business Review[12], as of way back in 1986, over fifty organizations had successfully used Scrum in *thousands* of projects to manage and control work, always with significant productivity improvements. Scrum, as illustrated in Figure 6.7, wraps an organization's existing engineering practices; they are improved as necessary while product increments are delivered to the user or marketplace. As is often said about Scrum, "oh, that's just my idea X by another name". Except Scrum is spelled out as values, practices, and rules in a development framework that can be quickly implemented and repeated.

---

[12] Harvard Business Review 86116:137-146, 1986

Figure 6.7 - Scrum is an agile process that allows PMs to focus on delivering the highest business value in the shortest time

As discussed in the earlier chapters of this book, scrum has been used to produce financial products, Internet products, and medical products. In every instance, the organization was log-jammed, unable to produce shippable products for such a long period of time that engineers, management, and investors were deeply concerned. Scrum broke the logjam and began incremental product delivery, often with the first shippable product occurring within the same quarter.

Trend: The APM Vision

The million dollar question is: Are you delivering innovative products to your customers?" Agile PMs should be always excited about developing innovative projects that stretch their limits and abilities as individuals and professionals, while creating a work environment in which people, as individuals and in teams, can thrive.

Keep in mind that without concrete practices, your principles are sterile; but without principles, practices have no life, no character, no heart. Although more than half of this book describes a lifecycle process and specific practices of APM, the other half is also very important—the half that attempts to articulate the values and principles behind the processes and practices. Avery successful agile PM will need daily convictions about the trade and expertise at hand.

As Jim Highsmith quoted in his book titled Agile Project Management: Creating Innovative Products[13],

> *The conviction at the core of the agile movement is creating a better workplace, free from tyranny, arbitrariness, and authoritarianism. Not free from structure. Not free from responsibility. Not free from project managers and executives making decisions. This conviction flows not only from wanting to create a progressive social architecture in which individuals can thrive, but also from the belief that agile social architectures produce the best, most innovative products. Where self-organization and self-discipline flourish, where processes are designed (and adapted) to support people rather than restrict them, where individual talents and skills are valued—great products emerge.*

[13] Addison Wesley, 2004, ISBN : 0-321-21977-5

References

Alistair Cockburn, Crystal Clear: A Human Powered Methodology for Small Teams

Bjorn Gustafsson, OpenUP - the best of two worlds

DSDM Public Version 4.2 Manual

Jim Highsmith, James A. Highsmith, Donald Eastlake, Kitty Niles, Agile software development ecosystems

Ken Schwaber, Agile Project Management with Scrum

Master's thesis on Agile Software development in theory and practice by Jonna Kalermo and Jenni Rissanen

Presentation on Feature driven development by Justin-Josef Angel

Ricardo Balduino, Introduction to OpenUP (Open Unified Process)

www.agilekiwi.com

www.ambysoft.com

www.bigvisible.com

www.ccpace.com

www.controlchaos.com

www.dsdm.org

www.featuredrivendevelopment.com

www.poppendieck.com

www.projectconnections.com

www.softwaremag.com

www.theserverside.com

www.wikipedia.com

www.xprogramming.com

About the Authors

About Marcus Goncalves

Marcus Goncalves has more than 20 years of management consulting experience in the U.S., Latin America, Europe, Middle East and Asia. Mr. Goncalves is the former CTO and CKO of Virtual Access Networks, which under his leadership, and project management skills, was awarded the *Best Enterprise Product* at Comdex Fall 2001, leading to the acquisition of the company by Symantec. He holds a Masters Degree in CIS, a BA in Business Administration, and is an Ed.D. Candidate at Boston University School of Education. He has more than 36 books published in the U.S. and many others in Brazil, Japan, China, Taiwan, Germany, Spain and Romania. He is often invited to speak on these subjects worldwide. Marcus is an Assistant Professor and the International Business Chair at Nichols College, and can be contacted via email at marcus.goncalves@nichols.edu or at marcusg@marcusgoncalves.com.

About Raj Heda, PMP@

Raj Heda has more than 12 years of work experience in various Information Technology areas, with broad consulting, leadership, teamwork and project management skills. He currently works as an Associate Project Manager with IBM. He has two years of experience teaching various IT and Management courses at Nichols College. He holds certifications for Project Management Professional (from PMI), IBM Lotus Notes 8.0, IBM Quickr 6 and IBM WebSphere Portal 6.0. He holds a Master's Degree in CIS from Boston University. He has one patent filed and five accepted for publication to his credit. He is the lead author for the upcoming book on Lotus Notes 8.5 to be published by IBM Press and Pearson publications. He is on the exam review board of THE Portal and Lotus Notes products of IBM. He can be reached at rheda@nichols.edu.

www.ingramcontent.com/pod-product-compliance
Lightning Source LLC
Chambersburg PA
CBHW060946050326
40689CB00012B/2571